Gemma Milne is a Scottish science and technology writer whose works have been featured in the BBC, the *Guardian*, *The Times*, *OneZero*, *Quartz* and others. She is also the Deep Tech and Science Startup Contributor for Forbes Europe. Gemma is the co-founder of Science: Disrupt – a media outlet covering advances in science startups, research processes and industries such as space, energy, health and advanced computing. She works with the World Economic Forum as one of their Global Shapers, and is also an advisor to the European Commission and Innovate UK, helping them decide which scientific innovations should be funded with government money. She is also a Venture Scout for venture capital firm Backed VC. She is on the Innovation Juries for SXSW and the International Academy of Digital Arts and Sciences, and has spoken at conferences worldwide, such as SXSW, TEDx, Slush, Cannes Lions and Startup Grind on many different science and technology topics.

SMOKE & MIRRORS

HOW HYPE *OBSCURES* THE FUTURE AND HOW *TO SEE* PAST IT

GEMMA MILNE

ROBINSON

ROBINSON

First published in Great Britain
in 2020 by Robinson

Copyright © Gemma Milne, 2020

1 3 5 7 9 10 8 6 4 2

A CIP catalogue record for this book is
available from the British Library

ISBN: 978-1-47214-366-2

Typeset in Scala by Hewer Text UK Ltd
Printed and bound in Great Britain
by Clays Ltd, Elcograf S.p.A.

Papers used by Robinson
are from well-managed forests
and other responsible sources

Robinson
An imprint of
Little, Brown Book Group
Carmelite House
50 Victoria Embankment
London EC4Y 0DZ

An Hachette UK Company
www.hachette.co.uk

www.littlebrown.co.uk

For Papa

Contents

Introduction

It started with ghosts in the 1700s. 'Conjured ghosts.'

A box would hide the projector, beaming a spooky image at an angled mirror inside, and the ghoulish form would emerge from the hole cut in top of the box to light up the plume of smoke generated on stage, to the horror of the audience apprehensively looking on.

Magicians would use the trick to delight audiences; charlatans would use it to con; but soon the smoke-and-mirror technique would become a classic in the magic toolbox, known as the original clever way of creating illusions.[1]

Nowadays, most magicians are known as entertainers, not witches and wizards. The audience walk into those theatres ready to suspend their disbelief in the pursuit of wonder and fun. Maybe one person wants to sit back and be wowed, maybe another wants to play detective to spot the sleight of hand, but either way, they know, and accept, that the magician on stage will be on a mission to fool.

And the magicians are masters of the craft of fooling. There are the physical props with hidden parts inside, but there's also their deep knowledge of how to create a trick in the mind of the audience. Magicians know exactly how to shift someone's thought process onto their desired path; to make someone believe something impossible; to misdirect. Without a degree of psychological manipulation, there is no magic trick.

Of course, magicians are honest deceivers. There's a reason that David Devant, the first president of the famous Magic Circle, is remembered for his catchphrase: 'all done with kindness'. And, of course, non-consensual fooling – or lying, as it's more commonly known – is almost never a truly kind act.

But what does it mean to be *accidentally* fooled? To be swept up in tantalising ideas, or trusted opinions, or simple hope that something false is true? To be tricked somewhat, not out of malice from a magician of sorts behind the scenes, but by simplified narrative, distracted focus or an unchecked desire for something to be true cloaking rational thought?

This powerful tool capable of misleading is hype. And it's powerful because it's not always a tool that someone somewhere has chosen to wield, but because it can so easily be misconstrued, carelessly used and spread like wildfire.

There are many stories of lying, deception and misuse of power; but I'd argue that hype can be even more damaging than deliberate, sometimes criminal, actions. Conspiracies too make for intriguing reading, but the curious power of hype is one that, in my mind, makes for an under-discussed truth much stranger and more compelling than speculation.

Before we go on, we should probably define hype. Some describe it as extravagant or intensive promotion; others define it as the use of a lot of advertising to make people interested in something; others go for, simply, exaggerated publicity. Whatever specific words you go for, it's clear that when we use 'hype' to describe sentiment, it is somewhat biased towards a particular perception not because of the *facts* about its merits, but because of the particular *words and narratives* surrounding it.

Sometimes those words and narratives are warranted; sometimes they are not. Sometimes hype perfectly encapsulates the future being built; sometimes it obscures it. Sometimes it results in socially beneficial over-embellishment; sometimes it changes the course of progress for the worse. The crucial thing then is not to halt all hype or banish its use across the board, but instead to understand when we are presented with hype and how to sort the fair from the problematic.

Hype, like any tool, isn't inherently good or bad. It can be the tool with which we gather communities around positive societal change, and it can be the tool that misleads to satisfy the ill-conceived wants of a few immoral actors. Sometimes people don't even know they're propagating it. But when hype starts to grow unchecked, it doesn't really matter who started it and why; what matters is that it is spotted before any damage happens.

Unfortunately, and in some cases, tragically, this can be far easier said than done.

Those who work in science and technology are often tasked with and driven by making the world a better place. And understandably so: their work is mostly comprised of progressing onwards from one state of knowledge to the next, and then applying this new knowledge to the world around them. It's broadly an ambitious activity, fuelled by optimism, tenacity and unwillingness to settle for an unsatisfactory result when more can still be done.

It's also a complex area, both in terms of depth of understanding of those at the coal face, as well as keeping up with the pace of discovery and invention. And in trying to work out exactly what's happening in science and technology, you can often be confronted with a lack of understandable information, an abundance of over-simplified generalisation and indecipherable ambiguity. It's in these conditions that hype tends to be wielded as a tool to aid communication and comprehension, but they're also the conditions that make hype ripe for unhinged propagation.

Beating hype individually can be as simple as spotting it, as its power is in its illusion, but if we're to avoid consequences that affect society at large, spotting and contextualising hype must be done collectively by the masses to stop it in its tracks.

My hope with this book is to show you that anyone, regardless of scientific background and confidence, can learn to see past hype. We can't get it right every time, but surely with more people even just stopping to consider whether something they hear, see or read is hype, our collective societal intelligence massively increases.

I'm a science and technology writer, so it's my job to meet incredible inventors, thinkers and entrepreneurs and tell their stories to the world.

It's also my job to make sure I'm writing thoughtfully, carefully and ethically, knowing that my words can have power in raising people and companies up, or curbing problematic actors. I also work with a venture capital fund, helping them find promising young startups, and I advise government on which company missions merit public money and public hope. My job means trying to get to the bottom of what's going on in industries, who is legitimate, and what the future potential looks like through many different lenses – financial, societal and more. I'm paid to think critically.

Before I became a writer, I worked in advertising, where I learned the power of messaging both for good and not for good. I watched big companies pick a narrative that fitted the zeitgeist, regardless of how authentic it was for the reality of their company, and sway consumers towards their stories and their stores. I also worked in corporate innovation in advertising, where my job was to travel the world to work out what the latest trends in technology and startups were for the agency's clients – and I learned that innovation can regularly be a hype-driven marketing tool much more so than a mindset or a research and development strategy.

In short, hype has played a huge role in my career, in finding the best ways to create it, in spotting it, in writing about it, and in striving to beat it. I've seen hype do amazing things, frustrate the hell out of me, and horrify me when I can clearly see the dangerous behaviour it can inspire.

I'm by no means perfect in spotting and beating hype – no one is, as we'll discover throughout this book – but I'm fortunate in that many of my jobs have had some element of hype attached to them, and the diversity of jobs I've done (we didn't even touch on the brief investment banking foray, the summer-camp chef, the door-to-door salesperson, or the wedding waitress) means I tend to look at things through various lenses automatically. I'm a jack of many trades, master of not-getting-stuck-in-a-silo, as it were.

I've chosen nine areas of science and technology through which to tell this story of hype. These specific areas are of interest because they are the 'moonshots' of our time, and thus they carry many different forms of hype within them. Some are regularly talked about in the popular media, some are only emerging out of the more insular science and technology world, others are beneath the surface. I've picked them as they all represent a different kind of hype, and each story of hype results in different impacts on the world. They are all very legitimate areas of science and technology; I've broadly stayed away from hype around pseudoscience and stuck to misguided hype around reality. I believe we should all, as citizens of the world, have a firmer grip on the areas covered in this book, regardless of our scientific background. Science and technology are part of society, after all, and being swept up in hype in these crucial areas means – as we'll discover – potentially very problematic effects on our planet and on people.

This book is split into three parts: Now, Next and Nearing. Within each part are three areas of science and technology relevant to that timeframe, the role that hype plays within them, the existing and possible future repercussions of that hype, and an exploration of how we might spot and contextualise that hype moving forward. Each role of hype, and each 'method' of spotting it, can arguably be applied across many chapters and indeed other areas of science and technology this book doesn't include.

The point is not to just take my word as gospel in these specific realms, but to show how we can all go about seeking out the patterns that hype exhibits and the tactics we can employ beyond these pages.

In the first part, Now, we'll see the current impact that hype has on our world, in how it helps retain a problematic status quo, how it can be a double-edged sword and how it can shield complexity. In Next, we'll see how hype is currently swaying the development of crucial fields in how it can curb action, how it can shield flaws and how it can act just like a placebo of sorts, telling us one thing but fuelling another. Finally, in Nearing, we'll explore how hype affects us as individuals and ultimately damages future progress, in its fanatical nature, in its relinquishing of responsibility, and in its halting of the most crucial activity of them all: critical thinking.

Throughout, we'll explore that which blinkers us from seeing past hype, such as marketing, perceived expertise, complexity, fear of being wrong, bad incentives, human psychology and idealism. We'll see both why they blinker us so easily, and what can be done to remove them as hurdles.

Spotting and contextualising hype isn't about learning the nitty-gritty of how science and technology work, but rather seeing how they fit into the systems of society and looking at ideas through various different perspectives, such as media, financial markets, law, geopolitics, socio-economics and the environment. No science and technology book is complete without a little intertwining of how the world actually works.

We'll also discuss those uncomfortable truths about our more active susceptibility to hype, our reluctance to change and our dissociation of various kinds of responsibility.

Hype tends to only rear its ugly head when the majority don't know that it's hype; when we don't collectively realise that we're consuming amped-up ideas as opposed to those celebrated on merit. And without being able to tell one from the other, what chance do we have actually to engage with science and technology? To consider whether we agree with the celebration of merit or not? To vote for a better future with our behaviour and our ballot papers armed with more, better information?

Hype is an effective tool, there's no doubt about that, but if we want to wield its power responsibly, or simply curb its potentially harmful effects, we must first work out how the sleight of hand is done.

And would you believe, I have the book of magic tricks right here.

PART ONE

Now

Finding the True Meaning of Value in the World of Farming

Have you ever considered how strangely common bananas are?

If, like me, you live in Europe, the US or some parts of Asia, you'll know that bananas aren't grown anywhere close to where we live, and yet they are a staple part of most of our diets. They are the first thing we see when we walk into a supermarket, showing the store off to be the perfect picture of health, freshness and abundance. They go into our lunchboxes and our healthy shakes and our baked loaves. We debate as to whether they brown faster in the fridge or on the countertop. Some people are freaked out by the stringy bits, some people prefer them more ripe, some people can't get their heads around why everyone else likes bananas so much.

Bananas are cheap and easy, and we love them for it.

There's a hidden cost behind bananas, though. In fact, there are several. From low wages, low income and lack of social security for those who work on and own small banana plantations to high environmental costs including huge land occupation, overuse of fertilisers and water depletion, the true cost of bananas is not at all reflected in the price we pay at the supermarket.[1] The true cost of bananas is felt in the effects of agriculture on our planet, our health, and the widening of the wealth gap. The price is paid, but not in the supermarket.

It's not just bananas, of course. Much of the food we see on supermarket shelves is priced as low as possible to drive customers into stores.

Some fruit and veg we buy is actually priced lower than what the supermarket paid for it, just to encourage us to choose one shop over another; they are the so-called 'loss-leaders'.[2] Think of the supermarket adverts leading with their competitive prices for the basics: 'a bunch of bananas for only 90p!', prompting us to choose that shop for our full weekly food purchase at the expense of a small loss over a common fruit.

Cheap food isn't just about getting customers into supermarkets. It's not just a cynical corporate move. We, the public, demand low prices; and when prices go up, we see it as a sign of our governments failing. Changing prices can quickly become political.

In Britain we spend only 9 per cent of our total household outgoings on our food shopping. After the US and Singapore, Britain comes third in the race to the bottom of spending on food.[3] We also spend less than our parents and grandparents did, with the proportion of our income dedicated to food more than halving since the 1950s.[4]

Despite our relatively cheap food, though, we balk at the idea of food prices going up, no matter the rationale. In 2018, a UK summer heatwave reduced feed for dairy cattle, hugely impacting dairy farmers, and the price of milk subsequently went up. Newspapers widely covered the understandable rise in price, with *The Express* describing it as 'soaring', and presented the general public as the party to lose out, the *Metro* writing: 'Shoppers could be facing paying as much as 7.5 per cent more for a pint of milk'.[5] In 2016, the reduced value of the British pound following the Brexit vote resulted in Unilever increasing their wholesale price of the UK's beloved (by some) Marmite. This then caused Tesco to refuse to stock it, and prompted the court of public opinion to brand the trade dispute 'Marmitegate'.[6]

When these price changes happen due to global market and environmental effects, we tend not to consider how cheap our food really is with respect to how it gets from the farm to our plate, and instead point the finger at the 'broken agriculture system', the government or greedy companies. Artificially low food prices ironically have some of the biggest negative repercussions for those at the lower end of the socioeconomic scale. The lowest prices are mostly preserved for the unhealthiest of food; the cheapest calories tend to be the least healthy.

Responsibility for the low price of food, and high costs hidden behind the scenes, doesn't sit with one player alone. But the party that arguably has one of the biggest influences in the way our food is priced – and thus, how farming is done – is the one we regularly omit in our complaints towards the food industry.

It's also the one we find hardest to face: ourselves.

PRICE VERSUS COST

The seemingly 'darker' stories of what goes on behind our food do make headlines. There are plenty of 'well, actually'-type articles telling vegans that opting for cashew milk is causing farm workers to be exposed to dangerous toxic materials, when often not provided with gloves.[7] Or there are the ones telling millennials that their avocado obsession is fuelling murderous Mexican drug cartels over the 'green gold' growing in the Michoacán region.[8] There's also been increasing coverage of the climate effects that farming, specifically dairy and meat farming, has on the planet, prompting much petitioning and public discussion about the benefits of vegan diets.

Of course, these stories aren't just inflammatory clickbait, as there are human rights violations that happen the world over in industries that ultimately make a lot of money; agriculture is not so different from the diamond trade, the electronics industry and the fast-fashion manufacturers in that regard. And for people of certain cultures, socioeconomic classes and personal situations, veganism or a refusal to buy certain foods from certain shops make sense with respect to individual values and health choices. For many, veganism is considered the only logical choice given the industrialisation of the meat and dairy industry.

But the food system is not just a machine in need of an upgrade. There are even bigger problems in the food industry, and they are far more nuanced than a consumer boycott can solve. The evidence that the food industry is contributing to negative societal, environmental and health impacts is piling up, and the reasons for these huge hidden costs are often left in research papers and out of the public consciousness, and thus are not at all factored into the price we pay for food.

The lack of understanding of the real inherent issues of the agricultural system also leads to naive optimism in the approaches to solving problems. There's the idealistic view that we should all 'go back to the glory days' of farming before it became industrialised, but that doesn't solve the problem of meeting demand. There is also a modern idealism that has crept in, where there are huge efforts to use innovations in science and technology to make agriculture more sustainable, more productive, cheaper and more 'on demand'. This modern idealism is leading many to believe the hype that 'technology will save us'. These efforts are useful to some extent, but they are Band-Aids on a gaping wound.

If we avoid buying some foods for environmental reasons without knowing if there will be poverty-inducing repercussions for farmers at the coal face, it only 'solves' one problem of the food system and exacerbates another. Similarly, we cannot invent our way out of our food system's failings; the system is far too complex and interdependent on social, environmental and economic issues for a series of high-tech Band-Aids.

Going vegan, eating less beef and buying from a local farmers' market are all positive actions, which many are understandably increasingly moving towards. And innovative technologies and scientific discoveries can go some way to solving some pieces of the agricultural puzzle. But without a shift in broader economic values, and an understanding that the price and real cost of some foods simply don't add up, individual and entrepreneurial purpose-driven actions run the risk of simply scratching the surface.

The food system doesn't need fixing, it needs redesign. It needs to be a system not fuelled by reduction in price, but by ensuring a sustainable, fair, healthy world.

Individual and entrepreneurial efforts are great at tackling specific issues, and they are the seeds of the shift in consumer values required for a new way of doing things, but we cannot give in to the hype that these are the only efforts required. A more fundamental change is necessary, and if we believe that tech and boycotts are all that's needed, we'll miss the opportunities for change when they present themselves.

In order to spot those opportunities, though, we need to understand the scale and range of the problems in the agriculture industry and work out how to carefully, quickly and without causing more damage rebuild this well-oiled but highly problematic agri-industrial machine with environmental and ethical concerns at its heart.

IT CAN'T CARRY ON LIKE THIS

When sustainability of agriculture is discussed, often greenhouse emissions are the main topic of conversation. Understandably so, considering the fact that the agriculture industry only tails the transport and electricity-generation industries in its total emissions produced. Of those emissions, cows are responsible for two-thirds, due to the methane that they produce.[9]

Sustainable farming means more than farming that doesn't create so many emissions; it means farming that doesn't deplete all our natural resources, and that can be maintained as society continues to develop. And this is where the bigger problems start to arise.

For example, agriculture is responsible for 70 per cent of all global freshwater withdrawals, reducing the availability of clean water for other industries and uses.[10] As the world warms with climate change, water also becomes more scarce. The agriculture industry not only uses most of the water, but the fertilisers and pesticides used in farming pollute the water in the surrounding area, having an even greater impact on clean water provision.[11] It's a particular combination of the chemical use and certain farming practices that leads to water contamination, and although there are farming techniques that keep the chemicals in the soil, they are more expensive and intensive and therefore less common. It's also worth pointing out that in some developing countries where there are no formal resources to aid farmers, there's little in the way of education for farmers to learn more sustainable practices.

Also, land use in agriculture is hugely problematic. Not only does farming take up around half of available liveable land, expansion of farms results in deforestation and destruction of precious animal ecosystems.[12] The land we farm on is also inefficient, with 77 per cent of farmland used for livestock, which supplies society with only 17 per cent of our calories.[13]

It's not just water and land that are needed to grow or nourish our food sources; plants, animals and micro-organisms act as natural support systems. There are the bees, insects, bats and birds that pollinate three-quarters of the world's crops but are drastically declining, some threatened with extinction.[14] There's the fact that our soil is becoming less productive due to its biodiversity being reduced drastically by pesticides, fertilisers and climate change. There's the severe reduction in diversity of plant species as we home in on plants that are cheaper and easier to farm; 60 per cent of the world's total crop production is only four species (maize, rice, wheat and soya), while the vast majority of the 7000 plants we are able to grow for produce are in steep decline.[15] There's the loss of green spaces to make way for concrete cities and roads. And the agriculture industry itself is to blame, too, with its expansion into wild land leaving less room for living things to thrive.

Food demand is also growing, not only due to population growth, but also urbanisation, global income growth and changing consumer choices. And it's not just about more food, but about fresher, healthier, trendier food, too. All this means that by 2050, if we stick with our current agricultural systems and processes, we'll need 65 per cent more water and 67 per cent more land, and will produce 87 per cent more greenhouse gases in our bid to satisfy both the hunger and demand for choice of the estimated 10 billion people that will be living on Earth.[16]

When we say that farming today is not sustainable, we don't just mean that the industry contributes to global warming: we mean that we simply don't have enough resources on Earth to satisfy future demand.

We cannot go on like this.

Our Food System Is Killing Us

Out of the 7 billion people who live on Earth now, 3 billion are plagued by malnutrition.[17] The number is so high because malnutrition doesn't just refer to those who are hungry from lack of food, it also refers to those who are obese and those who are deficient in micronutrients due to poor diet.

With increases in cheap highly processed food, obesity and micronutrient deficiency are prevalent worldwide. Climate change and conflict hugely impact food security across nations, particularly in the developing world. Nutrition is so bad, it was found that poor diets were responsible for 11 million deaths worldwide in 2017.[18] Tobacco, on the other hand, causes about 7 million deaths annually.[19] Malnutrition leads to increases in cancer, heart disease, diabetes and all sorts of other chronic illnesses, thus adding to the burden on national healthcare systems.

There are 800 million people worldwide who live in hunger. That's a hard number to accept considering a third of food produced for human consumption is wasted each year.[20] That's enough to feed 2 billion people.[21] In developing countries, this waste is usually down to poor logistics and transport infrastructure – a frustrating answer to the question of why people are starving.

Good nutrition is required to grow economies, and yet many efforts that focus on solving the world's nutrition problems sit firmly in the charitable sector.[22] Companies donate to malnutrition-mitigating efforts, but little is being done outside of the corporate social responsibility departments. The World Food Programme, set up to fight hunger worldwide, has a budget made up of governmental aid funds, corporate giving programmes and individual donors.[23] If companies and governments continue to see tackling nutrition as a charitable effort as opposed to an economic one, thus underfunding and undervaluing solutions, the malnutrition problem worldwide will continue to be too big to solve, will continue to kill people, and will continue to reduce the prosperity of nations.

And of those 800 million people who suffer from hunger worldwide, three-quarters of them are farmers.[24] It's a vicious irony that those who produce food have a lack of it to eat themselves.

Suicide is also rife within the farming profession. As climate change continues to cause unexpected adverse weather events worldwide, causing unpredictable and uncontrollable crop yield or destroying harvests altogether, farmers' income is reduced to the point they are going hungry or, in some cases, taking their own lives. In the US, the farmer suicide rate is almost double the national average. One farmer every four days takes their own life in Australia; in France, it's every two days. Farmers kill themselves in China to protest against urbanisation. India reports over 17,000 farmer suicides every year. The instability of income for farmers is often found to be related to the self-inflicted deaths: an unusually wet winter in Ireland in 2012 led to both trouble growing hay for feed as well as an increase in the suicide rate; the farmer suicide rate went up ten times in the UK during the 2001 foot-and-mouth disease outbreak during which the government required farmers to slaughter their animals.[25]

Our food system is killing us. Not only is it killing us through the prevalence and cheapness of unhealthy food, it's killing those of us who grow the food by propagating the environmental issues causing fatal uncertainty and stress.

THE COST OF LOW-PRICE FOOD TO THE POOR

When the food industry is proven to be unsustainable and detrimental to human health worldwide, the hidden costs of food start to emerge. To some, the answer is to raise the price of food to shift incentives within the industry towards producing better food, altering practices to be more environmentally friendly, and paying farmers more.

Of course, the answer cannot be so simple, especially given that wealth distribution is becoming more polarised, yet everyone's got to eat. In the UK, the richest 10 per cent spend about £90 per household each week on their food shop; the poorest 10 per cent spend less than £30, making food inequality rife in one of the wealthiest nations on Earth.[26]

The price of fresh, healthy food has risen relative to the price of highly processed unhealthy foods. In the UK, almost 1 million people live in so-called 'food deserts'. These are places where fresh, healthy food is largely inaccessible due to poor public transport links to big supermarkets, and a wealth of small corner shops or fast-food outlets both providing predominantly unhealthy processed 'junk food'.[27] In the US, over 23 million people live in food deserts.[28]

Pair all this with the fact that in first-world countries, more than half of the food waste happens in the home, and this carefree attitude towards food – no matter the quality – is often attributed to the low prices that we find in supermarkets.[29]

But in the eyes of those in poverty, the idea that food prices are too low relative to their hidden costs could be perceived as almost insulting, and so getting access to better food for those who desperately need it, while ensuring that the whole supply chain of that better food is compensated fairly, is a difficult task. Easy-to-manufacture food, created by companies that underpay and pressurise suppliers and growers, is cheaper to buy, and the bad health outcomes as a result of that are killing the poor.[30]

'TECHNOLOGY WILL SAVE US'

The issues of the current food system centre on sustainability, malnutrition and the socioeconomic divide. And as these problems have slowly been emerging in recent times, scientific and entrepreneurial efforts have been increasingly pointed towards solving these monumental global issues.

The year 2018, it seems, was a breakout period for 'agrifood tech' startups – startups creating new innovations for farmers, manufacturers, restaurants, retailers and consumers. A tidy $16.9 billion was invested into these new companies intent on disrupting the food system, an increase of 550 per cent across the six preceding years.[31] There is much optimism and excitement around the prospect of pointing new technologies towards optimising the traditional agriculture sector, and using the power of science to grow more food at a lower societal cost. Most of the companies in the space have purpose-driven missions and the startup founders take on the difficult task of building new enterprises in the hope of making the world a better place for all.

There are many different startups focusing on many different problems. They are focused on making people more healthy, or making farming more sustainable, or making good food more accessible, or making farming more profitable for farmers. They are asking questions such as: Does it get rid of the issue of weather unpredictability? Does it reduce our reliance on animals for protein? Does it increase yield for farmers?

Is it nicer to the planet? Does it make the food last longer, and hence reduce waste?

The size of the food industry, and the number of problems it presents, mean there are ample opportunities for the entrepreneurial-minded to start building. The perceived potential of startups in solving the problems of the food industry has led to a lot of surrounding hype, and a general feeling that we can invent our way out of the mess that is agriculture today.

A note: some people might look at the food industry today and see it as anything but a mess. Farms yield more than ever and thus more food is being produced than ever before, which for some, is the system's only job. I'd argue that this viewpoint is a narrow and short-sighted one, as the food industry clearly isn't only about producing large volumes of food, and that if we continue as we are today, the system is simply unsustainable. In short, it may not be a mess in the eyes of some today but, without change, it very soon will become abundantly clear even to those with that narrow production-focused view.

Planet-friendly Animal-free Protein

The effect that meat farming has on the environment gets a lot of press. For red meat in particular, there's the methane emitted by the animals, but the livestock industry in general requires a lot of land for grazing, making it the single most important driver of deforestation. There are also poorly enforced rules around managing slurry and manure, and the land, once used for livestock, is overgrazed to such an extent it cannot then be used for growing crops.

On the other hand, around 1 billion people work in livestock farming, and they are overwhelmingly concentrated in developing countries. Some of the world's poorest people depend on meat and livestock to make a living.

Despite the media coverage of rising veganism, global consumption of meat is growing.[32] In high-income countries such as those in Europe, Oceania and North America, we eat a lot of meat, but there's a relative plateau. In countries such as China, for instance, where income has been increasing for its citizens, eating meat comes with higher wages, and thus overall consumption is growing. India might be the only exception here, in terms of official data, but there seems to be evidence that more meat is being eaten there than is being reported, due to social and cultural pressures.[33] All in all, meat eating is rising, and that in turn is only going to increase the environmental damage that comes with it.

The numbers are worrying, and this has prompted much innovation in creating so-called 'alternative proteins' to somehow reduce and curb society's reliance on meat. We need enough protein in our diets, and many of us are drawn to eating meat both from a nutritional and taste perspective, so simply removing meat altogether without anything to fill the gap would not only cause public outcry, but increased malnutrition in certain societies.

With plant alternatives, of course, there's the option of simply eating them in their natural form – for example, eating lentils, beans or chickpeas. Or there's the option of creating a vegetarian version of a meat classic, such as a bean burger. But for people who love to eat meat, simply removing it from their diet isn't so simple, as there's a craving for something that looks, feels and tastes like meat.

The Yeast Feast

Fermenting yeast is by no means a new idea. For thousands of years, we've known how to harness yeast's power to create two of the most wonderful foodstuffs: bread and alcohol.

Nowadays, yeast is being harnessed for something else: plant-based meat.

California-based Impossible Foods is possibly the best-known startup in the plant-based 'tastes like meat' space, and using modern biotechnology techniques, it was able to discover which molecules are at the root of a burger's meaty taste and smell. It turned out this molecule was heme, an iron-carrying molecule associated with one of the key muscle proteins, myoglobin. Heme isn't just found in cow muscle, though, it's also found in lesser amounts in soy plant roots, and so the team at Impossible Foods genetically engineered yeast cells to contain the soy plant gene responsible for heme, and set about producing high quantities of this vegetarian meat-tasting molecule by fermenting the modified yeast. The heme is also responsible for binding iron, which then binds with oxygen, creating the red 'blood' that oozes out of the Impossible Burger patty, creating an even more meaty illusion.

Beyond Meat is the other big name in plant-based burgers, and instead of using soy plant protein, it chose pea protein, and refuses to use any genetically modified products. For some people, avoiding soy and GMO foods is important. Instead, Beyond Meat uses biotechnological techniques essentially to match the parts of the plant that are also found in meat, extracting them, and rearranging and assembling them into a more meat-like architecture.

The products are already on the market, and both companies enjoyed a lot of success in 2019 in particular, with Beyond Meat going public on the stock markets and Impossible Burgers teaming up with Burger King.

GROWING MUSCLE IN THE LAB

Making burgers from plants is one thing; making burgers from real meat that's not from an animal is another.

'Cultured meat' is meat that is made from muscle tissue grown in a lab. A small number of stem cells are extracted from animal muscle, and when encouraged along with a nutrient-rich serum, they grow. The meat is real meat, but opposed to growing inside the body of the animal, scientists essentially replicate that process outside the cow, pig, chicken or whichever animal you please. And, in theory, just one stem cell could be used to grow infinite amounts of meat, meaning the animals really are not required.

There are many startups in the space, and many animal products being cultured. There's Memphis Meats and Mosa Meats, Finless Foods focusing on fish, Perfect Day Foods focusing on dairy products, Clara Foods focusing on eggs; the list goes on.

In 2013, the first lab-grown burger was removed from its Petri dish, cooked in front of the world's press, and tasted by food critics – all to prove to the world that lab-grown meat was not only doable, but safe, edible and tasty. At the time, that burger took two years and $330,000 to create, but since then the costs have started to come down.[34]

Beyond cost, there's the issue of upscaling. A bioreactor is required to create the perfect conditions for the muscle cells to grow, and the largest ones out there can only create enough meat to feed ten thousand people in a year, meaning many of these would be needed to create an economical meat-growing plant. Another issue is the serum: the mixtures that are currently most successful require animal blood, meaning animals would indeed still be required until new serums have been invented. To do this, scientists need to work out exactly what it is in the blood, out of thousands of substances, that makes it so effective in the serum, and then find ways of artificially creating that specific substance without the animal.

BUGS FOR DINNER

Snack packs of crickets and mealworms have been around for a while, as a high protein gym snack or something those into experimental food might incorporate into niche dinner parties, and have been farmed as foods in many societies for thousands of years. In parts of central Africa, up to half of the protein required for a healthy diet has historically come from insects.[35]

Nowadays, the push towards finding alternative proteins means that snack packs of freeze-dried critters are becoming more common, but overcoming the 'ick' factor that comes with eating bugs is no mean feat in many cultures.

Over the last few years, though, much research and entrepreneurial activity have gone into working out how to farm insects at scale and turn them into flours, oils and other base ingredients, instead of expecting consumers not used to eating bugs to snack on crickets as opposed to beef jerky. Insects can also be used as feed for livestock, for instance for fish and poultry farming, replacing the very unsustainable fish meal as the feed of choice. One of the big benefits of insects being farmed at this more mass-market scale is that they can be farmed on feed that livestock cannot eat, and would otherwise have been wasted, such as the clean by-products of other industries, such as milling and alcohol production. Farming insects also looks to be far cheaper and far kinder to the environment relative to the proteins they can replace.

Work still needs to be done in finding ways to scale insect farming, while maintaining its environmental benefits, at a cost that makes it acceptable for those buying feed and base ingredients. Feed for fish farming in Europe can come in at only €1 per kilogram, whereas insect protein, without farming at scale, can cost around €100 per kilogram.[36] In 2019, though, French company Ÿnsect raised €100 million from investors to build the biggest automated insect farm in the world in northern France, but with this farm currently focused on feed, there's still work to be done in getting insects more directly into the food we eat.[37]

How to Sell Alternative Protein

The alternative protein market is booming, both rightly and understandably so, but it doesn't come without imperfections.

Ground beef is very cheap to buy, as a result of years of industrialised production, and this low price in some sense anchors expectations about what is a reasonable price to pay for the protein element of a meal. The plant-based burgers are more expensive to buy. Quorn, which is the one alternative protein that has managed to reach mass-market level to some degree, is still considered a niche vegetarian food many meat eaters aren't interested in buying. There's a risk that plant-based meat will have the same fate if consumer attitudes towards the price of food doesn't change.

Cultured meat, then, presents an interesting opportunity to convert meat eaters to alternative protein. The science is still currently far off, though, and the price so high that even as production methods come down, it could still be out of reach as a mass-market product. In 2019, there was a study that found that the CO_2 emissions from electricity generation in labs creating cultured meat could be more detrimental for the environment, in terms of greenhouse gases, than meat farming. This was based on the fact that methane isn't as damaging to the atmosphere compared to CO_2, as methane only sticks there for about twelve years, whereas CO_2 accumulates for millennia.[38]

Again, though, the biggest hurdle looks to be convincing *us* to make the switch.

Food decisions, of course, aren't just about price. For those who don't live below the poverty line and have more control over which foods are accessible to them, intrinsic beliefs, culture, personal taste, personal health and the media narratives that surround particular foods all play a role in helping us decide what to eat.[39]

What's particularly puzzling in the alternative meat space is the idea of the authenticity, or 'real-ness', of food. Cultured meat, despite being the same chemical make-up as the meat from animals, can easily be mistaken for a Frankenstein-esque science experiment. Plant-based meat can still feel 'not real' to some, and there has been controversy over the inputs in Beyond Meat in particular; in 2019, its stock dipped when a consumer group 'warned of chemicals in fake meat', as CNBC put it.[40] Quorn is considered 'not real food' by many people due to its number of additives, particularly in the context of modern 'clean-eating' health trends, despite its arguably more 'clean' effect on the environment. Balancing cleanliness with respect to health and cleanliness with respect to the planet might be a tough communication task. The push towards alternative protein is forcing us all to consider what 'real' food actually is, and what the costs and benefits are to the 'authentic food' demand.

LEARNING FROM PAST MISTAKES

Narratives that we assign and internalise aren't just based on personal choice and opinion, they are often swayed by misunderstanding, misinformation and an unwillingness to consider the possibility that what is already believed is incorrect. And the power of misinformed narrative hasn't been more overt than in the debates around GM (genetically modified) food.

GM food has had genes from other species inserted to give it desirable characteristics, such as not spoiling so quickly, growing larger or being resistant to certain diseases. There is scientific consensus that food derived from GM crops is no more of a threat to human health than non-GM food, but when these crops were introduced in the 1990s, there was much public debate and misunderstanding about whether they were safe.[41] Despite the American Medical Association, the National Academy of Sciences, the American Association for the Advancement of Science and the World Health Organization all taking a public stance that GM foods were safe, only slightly more than a third of US consumers believed them.[42] Words such as 'Frankenfish' are still used to describe GM food, and this anti-GM sentiment of the general public has fed into politics, with many countries around the world almost completely banning the use of GM crops for arguably the wrong reasons.[43]

GM food has been the topic of much debate over the years, and there are certainly good and bad sides to this technology as there are with any other, but some of the arguments against its use are rooted in misunderstanding of the science. (Some GM crops contribute to the monoculture of the agriculture industry, due to their need for certain pesticides. There are also issues around food companies owning the intellectual property of foods people depend on, which the biotech companies behind GM foods require.)

A study released in 2019, focused on the US, France and Germany, showed that the most extreme opponents of GM food know the least about science but believe they know the most.[44] The lead researcher, Philip Fernbach, said in an interview with the *Guardian*: 'If you don't know much, it's hard to assess how much you know . . . The feeling of understanding that they have then stops them from learning the truth.'[45] The study has parallels with a few key social psychology observations: the Dunning–Kruger effect, that those who are incompetent at something are unable to recognise their own incompetence; 'active information avoidance', where information that clashes with existing worldviews is avoided; and the 'backfire effect', the puzzling tendency of some people to become further entrenched in their own worldview after being presented with conflicting evidence.

In 2018, the European Court of Justice ruled that the latest advance in GM foods – editing genes, as opposed to inserting genes from other species, a much more precise process – would be subject to the same rules used for GM foods more broadly.[46] This decision was widely criticised by the European scientific community, claiming the decision would dramatically slow innovation in European crop science.[47] Some argued that the decision had been made based on misunderstanding, and a political will to side with the misinformed public.

If the alternative meat sector wants to avoid the damaging narratives that have so severely tarnished the GM-food field, communication around these new foods has to feed into how people think, with an understanding of what the existing narratives already are.

It's also worth pointing out that meat isn't eaten just to satisfy our protein need. Many of us eat for pleasure, for health, for community. If those trying to switch people away from environmentally damaging eating habits fail to also keep this in mind, adoption of alternative meat is likely to stay niche and broadly ineffective in solving the problems of the agriculture industry.

DIGITISING THE FARM

By using the power of the internet, sensors, mobile phones, communication networks, knowledge sharing and more intelligent analysis of data, some level of control can be gained over the unpredictable, manual labour-heavy, interdependent and complicated task of growing food well.

There's the technology that reduces the human labour required to run a farm, such as robotics and drones for managing crops and the barn-mounted facial recognition cameras capturing health information about livestock without having to check each animal individually. There's the technology that gives farmers more information on what they should farm and when, such as sensors tracking the soil, satellite data providing climate insights, and online marketplaces where buyers and farmers can connect their supply and demand, just like Airbnb connects travellers and available rooms. There's even the technology that allows for entirely different farming systems altogether, such as the indoor farms, and farms automated in their entirety.

It is estimated that there are more than 570 million farms worldwide and, of those, around 90 per cent are run by an individual farmer or a family and rely primarily on family labour. The vast majority of farms are small, many of which are less than two hectares in size, called smallholder farms.

And despite these farms having little space to work with, operating on only 12 per cent of the world's land available for agriculture, they produce 80 per cent of the food consumed in Asia and sub-Saharan Africa.[48]

Small and family-run farms, though, tend to be hit hardest by the things that make farming hard, such as unpredictable weather, crop disease, pests and fluctuating market prices. They don't have big companies underwriting their efforts, as large industrial farms do, and they rely on the transfer of knowledge from generation to generation to keep the farms in business. The average age of a farmer is sixty years old, and farming is becoming less attractive as an occupation due to increasingly poor farming conditions, both economically and environmentally.[49] Saying that, it's found that when younger farmers take an active role in family farms, new technology is more readily adopted and the income of those farms indeed goes up.[50]

Many of the digital innovations in farming have the potential to revolutionise farms all over the world, making small and family-run farms easier to manage, more profitable and more efficient. But many of the initial target customers for these new technologies and business models are the wealthy, technologically savvy industrial outfits, reflected in their high costs and effectiveness only at scale. And the digital innovations specifically targeted towards farmers in developing countries, where relatively simple technologies can have a huge impact, don't garner as much hype, investment and business support. The startups in this space are not so-called 'unicorns', so they're not creating huge sums of money in return on investment for funders, and some of the most impactful use-cases are simple applications of older, arguably more boring, technology, such as weather reports sent by text.

There seems to be a disconnect between what gets the most hype in the 'future farming' space versus what is hugely impactful for the majority of the world's farmers, and the people they are responsible for feeding.

There's one agritech innovation that epitomises this disconnect more than most, and that is the vertical farm.

THE RISE OF VERTICAL FARMING

In 2010, Dickson Despommier, an emeritus professor of microbiology and public health at Columbia University, published a book called *The Vertical Farm: Feeding the World in the 21st Century*. In it, he outlined a vision for the future of the agriculture industry, where food is grown in urban skyscrapers, using artificial lights, heaters, water pumps and the power of computers to grow crops indoors. He wrote about how these vertical farms would integrate food production into cities, increase the amount of produce per unit of land area, recycle the small amount of water required, protect crops from pests without needing to use pesticides, and reduce the distance food is transported to reach urban customers. His rationale behind the need for change was on point: he argued that soil is being exploited, that farming is inefficient, that the food system is broken.

The problem, though, is that vertical farming doesn't solve those problems.

Still, since the beginning of the century, the idea of growing food vertically in urban areas has increasingly captured the minds of architects, of engineers, of designers, of plucky entrepreneurs, and wealthy Silicon Valley investors. There are many startups out there that have created indoor systems, glowing magenta from the LEDs inside providing optimal light for photosynthesis, with trays of greenery line by line, stacked one on top of the other.

Of course, to the cannabis dealers in high-rise buildings, the idea might not seem quite so innovative, but the technology used in these indoor systems certainly is advanced. Whether it's in the converted shipping containers of Boston startup Freight Farms, or the huge warehouses of New Jersey startup Aerofarms, the advancement of hydroponics – growing plants using nutrient solutions as opposed to soil – and pairing that with intelligent computer systems monitoring and then optimising the growth, have resulted in technologically advanced ways of growing food.

Technologically advanced ways, however, are not always the best ways.

THE DOWNSIDE OF VERTICAL FARMING

For starters, the energy requirements of vertical farms are tremendous. For every square metre of lettuce-growing area, a traditionally heated greenhouse needs around 250 kWh of energy a year, whereas a vertical farm needs around 3500 kWh per year, with 98 per cent of that energy use due to artificial lighting and climate control.[51] Of course, renewable energy sources could be used, but the idea of taking some of the tiny amount of renewable energy which society harvests, some of which we get from the sun via solar panels, and then use it to replicate the sun in an indoor farm, begs the question of why we don't just use the sun in the first place.

Proponents of vertical farming will argue that because the food is grown closer to wherever it is consumed, there are fewer 'food miles' between the farm and table and thus less fuel emissions and greenhouse gases.

But reducing the number of miles food travels doesn't actually help the environment all that much, as local food typically requires the same amount of energy per pound to be transported as food from far away.[52] This is because food travelling long distances on ships, trains and lorries on motorways use less fuel, per pound per mile, than small trucks driving around cities. And the emissions from food miles pale into insignificance when you compare them to emissions from deforestation and methane emitted from cows. More local food systems are good for the environment but only when the farms themselves aren't using up huge amounts of energy to run them. The energy requirements for vertical farming essentially cancel out the benefits of them being local in the first place.

Then there are the costs involved.

Just one container from Freight Farms costs $104,000, plus shipping. You could get over ten acres of farmland for the same price in some parts of the US. And the produce is expensive too. The mini-lettuce that Green Line Growers sells costs more than double the typical price of organic lettuce sold in shops.[53]

For cannabis growers, the cost of running an indoor operation is worth it due to its high market price. Lettuce doesn't quite add up in the same way.

The target audience for this kind of farming, both in terms of those doing the farming and those buying from them, is clearly the affluent. Personally, I'm not sure we really need more gourmet lettuce.

And speaking of lettuce, that's one of very few crops that can actually be grown in an indoor vertical farm. The artificial conditions don't allow for the proper mix of heat and light to trigger all plant development stages, for instance, the stage at which a plant produces fruit. Vertical farms have a hard time growing anything beyond lettuce, leafy greens, herbs and edible flowers. If the technology does improve to the extent that potatoes, tomatoes and green beans, for example, were able to be grown in these farms, a lot of space would then be taken up by their inedible leaves, roots and stems.

Moreover, vegetables as a whole only occupy 3 per cent of US farmland anyway; in terms of feeding the nation, more leafy greens grown inside isn't really what the vast majority of people are asking for, or indeed need.[54] Salad is not the same as food; we still need all the farms outside the cities, growing food in the usual ways. Vertical farms don't reduce any of the bad practices still plaguing the agriculture industry; it's an add-on that also adds more to the burden of emissions.

Vertical Farming Still Excites

These issues with vertical farming are not hidden away or covered up as such, and yet the hype around vertical farms continues to propagate. The projections that the vertical farming market will hit $4 billion in 2020 keeps investors investing.[55] The headlines telling us that vertical farming will 'feed the world' keep the rest of us idealistic and hopeful.

The point we're missing when we are swept up in the vertical farming hype is that it is simply irrelevant to the lives of those living and working in the rural regions where the vast majority of our food is farmed.

And even when we do accept that it's a solution for the wealthy, we're caught up in our sentimental attachment to local food production without considering the broader environmental cost that comes with the vertical farming approach. There are many other innovative local farming systems, such as rooftop aquaponics that use fish excrement to grow in the sun, or 'green-walls' that grow crops vertically outside, which use far less energy as they don't require artificial lighting.[56] In the UK city of Manchester, there are 136 hectares' worth of flat unoccupied rooftops.[57] Is our sentimentality and attachment to vertical farming therefore simply trendy?

When you consider the entire farming system, vertical farming does nothing to end soil abuse as all the soil farms are still required; it does little to improve malnutrition worldwide (or even within the cities in which the farms reside) due to the high cost of produce; it does nothing to make the lives of the small-scale farmers, operating 90 per cent of the world's farms, any easier.

There's an argument to be made that the investment of over $150 million into vertical farming startups in 2018 could have been put to better use, if the ultimate goal is to improve the system as a whole, as opposed to create high-priced products for the few.[58]

Technology (Alone) Can't Save Us

There are many incredible efforts happening in the agritech space that have the potential to have a real effect on farmers and consumers both now and in the future. The excitement in the space is fair, but only to some degree.

First, the majority of the investment and attention is focused on solutions that still play into the overarching problems of the agriculture industry. The biggest piece of the total investment into agritech in 2018, $3.9 billion worth, went towards restaurant marketplaces and food delivery companies. The second biggest, $3.6 billion, went to online food stores. Adding them together means that 44 per cent of the total amount of money put towards trying to make change to the global agriculture sector went towards door-to-door food delivery.[59]

Second, there's a limit as to how much impact the new innovations can have in solving the biggest food industry problems. The plant-based burgers and the vegan milk products might be able to chip away at some of the problematic agricultural practices in developed countries, but they do little for food insecurity and global malnutrition if not done effectively at scale. Digital solutions that are only affordable or applicable to large industrial farms have only marginal effects in improving the lives and livelihoods of farmers around the globe.

When thinking about innovation in science and technology with respect to the food and agriculture industry, you have to consider it in two halves: there are those who are making food better for those who can afford it, and those making food and farming better for those who can't. Much of the hype surrounding agritech uses narratives related to the latter group to sell innovations that sit firmly in the former. Vertical farms and other efforts targeted towards the wealthy simply shouldn't be spoken about in the context of 'feeding the world' or 'revolutionising agriculture'.

There's nothing wrong with choosing high-income customers as the target audience, but using 'save the world' narratives to sell your vision is not only misleading, it does nothing to change the broader agriculture-industry status quo.

Not all innovators and entrepreneurs are deliberately using these narratives to hype up their companies and inventions. But there is a severe lack of discussion, acceptance and vocal challenging of the elephant in the room that affects each and every person operating in the industry, which projects a kind of 'we can just invent our way out of this mess' idealism that is simply unrealistic.

That elephant, which incentivises problematic behaviour, puts value on the cheapness of food as opposed to the environmental, health and social impact, and creates an ever-expanding difference in wealth between large agribusinesses and small family farms, is the government subsidy.

SUBSTANDARD SUBSIDIES

Agricultural subsidies are payments and other kinds of support that a government makes to farmers for producing enough food to feed the nation. The subsidies are meant to protect farmers from unexpected economic disasters (the US programme began in the 1930s during the Great Depression), reduce the impact of unpredictable weather on farmer income, and generally support the industry that provides argu-ably the most important thing citizens need.

Subsidies sound good in theory but the reality is that, in many countries, subsidies do far more harm, and actually are what are keeping the flawed agricultural industry from innovating, shifting to more sustainable practices and improving the health of society.

The US government currently gives out about $25 billion in farm subsidies annually; in Europe, the CAP (Common Agricultural Policy) gives out €50 billion per year to the farmers across the continent. Many people believe that subsidies are predominantly there to support small family farms, but in both the US and in Europe, the vast majority of subsidies are funnelled to industrial scale agribusinesses and large farms.

In Europe, this is mostly due to the fact that the subsidies are allocated based on the amount of an owner's land that can be farmed. And that land doesn't even *need* to be farmed, as the rules only stipulate that the land has to be ready at the owner's disposal.[60] You cannot claim subsidies unless you own or lease at least five hectares, meaning small farms are entirely ineligible.[61] Even within the recipient pool though, there's a massive skew towards giving huge lump sums to the few wealthy farms and, at the other end of the scale, smaller amounts to the many (still quite large) farms. The top 10 per cent of recipients receive nearly half the total subsidy, while the bottom 20 per cent receive only 2 per cent.[62]

In the US it's a similar story, with the top 10 per cent of farm businesses receiving 78 per cent of the subsidy.[63] Most farms don't receive anything at all, with only thirty per cent of US farms eligible for the payments for their produce, and only 16 per cent of farms able to take part in the government's crop-insurance programme.[64] Fifty people on the *Forbes* 400 list of the wealthiest Americans received farm subsidies, begging the question, why on earth do they need what are essentially welfare benefits?[65]

The objective injustice in the way subsidies are split between farms is only one side of the coin, though. The bigger problem with subsidies beyond being unfair is that they actively encourage the problematic behaviour that is at the root of almost all of the inherent issues of the agricultural industry.

Subsidies Make Farming Unsustainable

Back to Europe, where the CAP stipulates that subsidies are given to those who own land which can, in theory or in practice, be used for farming. The 'in theory' part here is the problem, as this means that there is land in Europe not being used for farming, but has been cleared of trees, wildlife and biodiversity in order to provide the right conditions for farming and take portions of taxpayer money. The UK version of the rules states that the land has to be free of so-called 'ineligible features', incentivising landowners to destroy wildlife habitats and create the perfect conditions for flooding by removing ponds, wide hedges, regenerating woodland or thriving salt marshes.[66]

In the US, out of all the crops that farmers grow, almost all of the subsidies go to just five: corn, soybeans, wheat, cotton and rice. Between 1995 and 2019, 93 per cent of US subsidies went to just those five crops.[67] This means that farmers who want public money to fund their business have to focus their farming on a single crop, and produce as much of it as can be squeezed out of the land. Environment-damaging fertilisers and pesticides are used to optimise the yield, crops aren't rotated (crop rotation helps to stop the soil from eroding), the soil is exploited and has a reduced ability to store and cycle carbon, and the surrounding water supplies are more polluted.

Almost 40 per cent of the corn produced in the US is for animal feed, another 40 per cent is for ethanol.[68] Subsidies for Texan cotton are close to $3 billion per year, and most of that cotton is shipped to China to make cheap clothes sold in US stores.[69] Of course, this is not the farmers' fault, they are simply growing what the market demands and pays them for. But the fact remains that public money, which is meant to be there to support the farmers who are feeding the nation, is being funnelled into big, profitable businesses that are not always in the business of feeding humans.

Producers of fruit and vegetables, on the other hand, are almost entirely missed out of the subsidy system.

Subsidies Make Farming Unhealthy

In 2016, it was found that this US subsidy focus on corn, wheat, soybeans, rice, cotton and dairy was having huge negative effects on public health.[70] The crops and livestock aren't necessarily unhealthy in their raw forms, but much of the US produce is turned into unhealthy products such as processed fatty foods, high fructose corn syrup and refined grains.

The study found that more than half of Americans' calories came from foods which were subsidised; that younger, poorer and less educated people ate far more subsidised foods; and the people who ate the most subsidised food had a 37 per cent higher risk of obesity, a 41 per cent greater risk of belly fat, and a 14 per cent higher risk of abnormal cholesterol.

The US government dietary guidelines tell Americans to avoid the same foods it makes cheap through subsidies.[71]

Subsidies Make Farming Unintelligent

North Dakota farmer Gabe Brown is a huge advocate for so-called regenerative farming, which, as the name might suggest, is a farming method that regenerates the soil that has been degraded by the traditional and industrial agriculture system. In his 2018 book *Dirt to Soil* he explains how one summer in North Dakota the weather shifted from several days of heat to freezing temperatures.[72] That shift took farmers by surprise, as hay and land for foraging was greatly reduced, leaving animals with not enough to eat. Some farmers travelled hundreds of miles to buy hay; others sold off their cattle. The government made a disaster declaration and started encouraging farmers to sign up for financial assistance.

Brown was encouraged to apply for the funding but he'd not really felt the effects of the changing weather. His farming method meant he could easily adapt his grazing strategy, and the years of regenerative farming meant his pastures were resilient to the swings in the weather. He wrote: 'I could have collected tens of thousands of dollars simply by walking into the Farm Service Agency office and signing my name on a form. But I just could not morally or ethically do that. We simply did not have a disaster on our operation.'[73] He said he saw what happened that summer as less of a natural disaster and rather a human-made one resulting from poor farm management.

But Can Sustainable Farming Pay?

Brown, and a growing movement of sustainable farming activists, have contributed to a fast-growing grassroots movement around regenerative agriculture. There are ample documentaries, farmer profiles and Facebook groups sharing stories of those on the ground who have shunned the subsidy system, and found ways to make their sustainable farming pay, while not costing the Earth in the process.

Regenerative agriculture is a catch-all term for farming that improves the land, mainly through rebuilding soil organic matter, which is plant or animal tissue in the process of decay. It only makes up between 2 per cent and 10 per cent of most soil, but for every 1 per cent increase in soil organic matter, the soil will hold 20,000 more gallons of water per acre.[74] That extra capacity to hold water means crops are more resilient when drought hits. More soil organic matter also means more nutritious soil, which leads to better protection against disease and pests. In short, better soil means more efficient farming, with less need for fertilisers, pesticides and other yield-squeezing chemicals.

The regenerative farming process includes integrating livestock and crop farming into the same areas, keeping soil covered at all times with plant residues and cover crops, no-till farming (which means not turning over the soil before planting new crops, to reduce the break-up of good soil structures), and diversifying the crops planted and managed on the farm. In short, the farmers farm in a more diverse way, meaning they can't just focus on one crop and optimise their subsidy payout in the process.

In 2019, Jessica McKenzie at the New Food Economy newsroom reported on the way some regenerative farming practices were actively discouraged by the US government.[75] She found many examples of farmers who were denied crop insurance payouts when big weather disasters struck, due to them not following the strict guidelines that would make them eligible. And even when the guidelines changed, many farmers still had to fight to get the payouts they were due, because of their perceived 'off book' farming methods. Even though crop insurance is administered by private companies, the government sets the rules, and if these rules are complicated to navigate, farmers are put off trying new methods despite the potentially huge benefits available from changing their practices. As McKenzie put it: 'If farmers want crop insurance, they have to play by the rules.'

The rules put farmers off experimenting with their own farm, from learning about more modern farming methods and trying different approaches, and generally taking control of how they want to run their farm.

If subsidies, including those that are given as insurance payouts, not only allowed but encouraged good soil maintenance combined with openness in supporting farmers in learning new, better, modern methods of adaptive farming, then there's a chance farms would be more resilient in times such as that summer Gabe Brown describes. And with more resilient farms, the need for subsidies might then decrease, and everyone in the system saves money rather than paying for the mistakes of the past.

Ban Subsidies?

Farm subsidies have been removed from entire countries before, including in New Zealand in 1984 as a result of a budget crisis. There was an uproar at the time, but over the decades New Zealand has continued to be a vital agricultural exporter, with farmers focused on satisfying the demands of the market as opposed to growing what is stipulated by government.[76] And the environmental effects have not been insignificant. The Federated Farmers of New Zealand reported better water quality, more effective land use and other benefits.[77]

Every country has different agricultural systems, though, and for some subsidies not only make sense, but are deemed necessary to keep the industry afloat. If we're to keep subsidies, however, with their vast influence and distortion of the agricultural market, we might as well distort the industry towards better environmental, socioeconomic and health-related activity.

For example, surely subsidies could be designed in a way that rewards farming that is environmentally kind, or farming that promotes biodiversity, or heals the soil, or pays more farmers across the socioeconomic spectrum (thus increasing purchasing power and spend in other industries), or improves public health. This might cost more in terms of upfront investment but, over the long term, investing in the land and the health of the nation is surely a cost-saving activity.

Either way, the discussion around subsidies needs to be ever louder, and more aligned to the values of the public as a whole, as opposed to simply that of the current measure of economic gain – the low price of food.

A Twentieth-century System

The agriculture industry, as it stands now, is entirely unsustainable; it is destroying our planet's precious resources, causing worldwide malnutrition, leading to suffering for farmers and contributing to mass food inequality worldwide. The price of food in the supermarket is too low in comparison with the hidden costs behind it and we are paying in healthcare costs, in a loss of the natural resources of our planet and in lack of compensation for those working on farms. But raising prices can further exacerbate the wealth distribution and health gaps, and it's a bad political move when consumers are of the belief that food should be priced cheaply. There are so many parts of the system that are now understood to not work, meaning there are almost infinite opportunities for incremental innovative solutions from startups and innovators, but the truth is that the system is so interdependent that any one solution most likely creates another problem elsewhere, or simply scratches the surface.

And all of this is built into a well-oiled system, incentivised and protected by the structure of government subsidies.

We have a twentieth-century agricultural system that is not fit for the realities of the twenty-first century. A system built to produce as much food as possible, as cheaply as possible, to provide urban wealthy consumers with whatever food they want when they want; and it is killing the planet and those who inhabit it.

But the system we have isn't broken, it just runs on the wrong values. It's a system where low price alone – instead of a healthy population, a healthy planet and a fair society – is equated with high value. We need a shift in what's perceived to be valuable by governments funding and regulating the system; a shift in what's perceived to be valuable by companies operating at all points in the chain; and a shift in our propensity to pay, in money and in behaviour, for the values we say we believe in.

What Is Value?

Economist Mariana Mazzucato thinks we've got the economic definition of value all wrong. If we track how value was defined through the ages, economists have focused on production of value from land, labour and capital, and the inherent value of something determined the price set for it. Nowadays, though, price is determined by supply and demand of the markets, and thus only that which manages to attract a high price is deemed to have value, regardless of how 'useful' it is to society. It doesn't matter whether that thing is a piece of nutritious food or a digital promise to buy something in ten years' time, the price it garners is whatever someone is willing to pay or willing to sell for. Value, then, is in the eye of the beholder.

Mazzucato believes this is hugely problematic as, in this set-up, we fail to account for whether something is actually productive or not for society as a whole. All we care about is the price attached to it. And that failing of the system to reward societal value with monetary value is linked to the way we measure our countries.

The international measure for the overall economic activity of a nation is GDP (Gross Domestic Product). It's used to determine how well a country is doing in terms of its economy's size and rate of growth, and since the end of the Second World War, promoting GDP growth has been the main national policy goal in almost every country.[78]

The problem with GDP, though, is that only goods and services sold in markets count. In Mazzucato's 2018 book *The Value of Everything*, she points out that because crime demands more police and security devices, more crime equals greater GDP. And along with crime, serious illness, hurricanes, divorce and pollution, with the services and products required to manage them and their aftermath, all contribute positively to total national GDP. The measure doesn't take into account effects on the environment, on social costs and on citizen health. GDP, simply, measures 'value' in an irrational way, and when countries obsess over raising their scores, governments miss what value really means. Mazzucato wants fairness, the environment, health, community and quality of life to be at the heart of economics.[79]

Other measures have been created, for example, the GPI (Genuine Progress Indicator), which takes into account many additional measures such as the value of volunteer work, the costs of crime and pollution, and income distribution.[80] Funnily enough, when GDP and GPI were compared across seventeen countries in 2013, the study found that the measures both rose in tandem from 1950 to around 1980, at which point GDP continued to climb and GPI flattened. By the end of the study in 2013, GPI was lower than in 1980, showing that the environmental and social costs were outweighing the benefits of increased GDP.[81]

THE VALUE OF FOOD

When market price and value are equated, we get a world which is unfair, unsustainable and unhealthy. The only way out of this is to open up the debate within sectors about what value really means.

For the food sector, that means creating a system that values the environment, societal health, socioeconomic stability and fairness above all else. And the beauty of this task is that we already know what that system can look like. We know that it means creating a subsidy system that rewards sustainable farming. We know that it means investing more into technology solutions that can plug the gap between the system we have now and a system fit for the twenty-first century.

And we know that it means shifting our own behaviour and expectations around food, and understanding better what good food really means in the context of a better fed, fairer and environmentally conscious society.

Tim Lang, professor of food policy at City University London, along with public health nutritionist Dr Pamela Mason, have defined what a sustainable diet looks like: increased intake of plant-based foods, with limited meat and processed foods high in sugar.[82] And they propose a thirty-year 'Great Transition' for food systems, optimised around six major overarching themes: food quality, health, environment, culture, economy and governance.

The thing stopping this from happening is a lack of momentum. When the scale of the problem and the underlying issues in the industry aren't clear, the public can only go by what's in the mainstream media. There's then no petition for change as spotting better food policies put forward by politicians is nigh-on impossible, meaning they are lost in the noise. Momentum doesn't gather around the ideas that need it the most.

And behaviour change at scale can only happen if the system surrounding us allows for an easier transition, and is clear in what it values. There's confusion about what to do as individuals and, with the cost of dietary change being so high, we want to know what's right before we do it. There are so many conflicting narratives surrounding our diets and cultural norms because they are always in the context of an out-of-date food system.

Expecting consumers to change without a functioning value-based system to adopt is unrealistic. Expecting entrepreneurs and inventors to pull some solution out of their magic science hat is to fundamentally misunderstand what's wrong with agriculture.

We Have to Do the Hard Thing

If we want the agriculture industry to change – which really means, if we want society to be healthy, fair and able to live well on planet Earth in the future – we need to accept that the solution is simple yet hard to achieve.

Shifting values is not an easy task. We don't want to come to terms with our personal problematic behaviour around food. We don't want to have to redesign entire business models around sustainable farming, losing short-term profits in the meantime. We don't want to have to rethink nationwide policies on how we subsidise farms and deal with the political damage that comes with big disruptive change when we can't see into the future and know for sure if it's the right thing to do.

We want the easy answer. We want to invent our way out of it. We want to be excited by the hype surrounding the latest and greatest technologies that are touted to solve our problems for us.

If we truly want to live in a better society, though, we have to accept that this desire must trump the others, and that what we all know deep down, but are refusing to believe, really is the only way forward: that we must accept the higher cost of better food, both in cash and behaviour change.

Change is hard; hype lets us momentarily question that truth. But as we continue to question the reality of the task in changing the problematic status quo, a better world only continues to move further out of our reach.

The Quest to Treat What We All Fear the Most

....

'A Cure for Cancer? Israeli Scientists May Have Found One', read the headline in the *Jerusalem Post* in January 2019. 'We'll Have a Cure for Cancer Within a Year', read the *New York Post* 5000 miles away.[1]

The scientists at the biotechnology company AEBi were the source of the story. They claimed to have found a cure for all forms of cancer and that they would make it available in only twelve months' time. AEBi's chairman, Dan Aridor, said: 'Our cancer cure will be effective from day one, will last a duration of a few weeks and will have no or minimal side effects at a much lower cost than most other treatments on the market.' He added: 'Our solution will be both generic and personal.'

It was news that any cancer patient, relative or anyone with even an inkling of an idea of how horrific cancer can be would rejoice at. The hope that those headlines proclaimed gained profit from elated clicks.

With vague evidence accompanying the articles, gained via the company website, it wasn't long before scientists the world over called out the irresponsible reporting.[2] The *Jerusalem Post* mentioned that the company's claims were based on a 'first exploratory mice experiment', as opposed to clinical trials in humans, in the second to last paragraph of its 1500-word piece; the *New York Post* didn't even mention it. Neither piece clarified that a universal cure was highly unlikely considering there are over two hundred types of cancer, which are each highly personalised to each individual. Neither article explained that taking a drug to market routinely takes over ten years.[3]

Neither article took into account the damaging effect unsubstantiated, bombastic claims have on those suffering with the disease, or on those who will get it in the years to come.

Science can be unpredictable sometimes, and huge discoveries can come as a surprise to the broader field. But that doesn't mean that coverage should make claims that haven't yet been proven. The headlines would have been more appropriate if they had come alongside real evidence and the actual availability of a working universal cancer cure. It would then surely be hailed as the most significant medical breakthrough in history, would invalidate much of what we already know about cancer and its exponential variability, and revolutionise the entire global healthcare system forever.

Except, that's not what happened, and it's not what is likely to happen either. The global media coverage told us otherwise.

A Familiar Yarn

For anyone who reads a tabloid that has a bustling health section, the cancer prevention stories won't be anything new. Calling out the conflicting pieces appearing mere months apart has become a bit of a morbid joke: 'Drinking coffee every day prevents cancer' versus 'Coffee is cancer's biggest ally'.

Stories about researchers and companies working in the cancer treatment space, however, make claims with grandiose language: they have found a key 'breakthrough', a 'game changing discovery' or a 'revolutionary method' in the 'battle to cure cancer'.[4] And these treatment stories appear across the entire media landscape.

The coverage of research, trials and patient 'miracles' is tantalising; the stories bring us a sense of collective positivity, and prompt both charitable donations and huge corporate cash injections for those working at the forefront of research. When the stories are clearly hyped up by the protagonists, or the headlines are clearly misrepresenting the research, the scientists will call them out, but those rebuttals are almost never as loud as the initial over-claim. And a little overhyping is also seen as allowable if the story is positive and the target is something so horrific as cancer: 'Come on, it might be early but they're just excited about going in the right direction,' might be the explanation behind the hype. Optimism and cheerleading are encouraged when the alternative can feel defeatist.

Hype in cancer research, though, is problematic. There's a lack of understanding of what cancer really is, and what it takes to manage and control the disease, meaning our ability to decode what is actually happening behind the headlines is nigh-on impossible. The warlike narratives of 'battling the alien within' and 'fighting for a cure' misrepresent the reality of those in the throes of experiencing or treating the disease. The relentless media coverage of even the most incremental step forward, as well as the triumphant slogans such as 'we will beat cancer', trick us into thinking society is further ahead than we actually are, and that a universal cure is both highly possible and soon to be found.

Many of the stories are so incredibly premature, and the narratives so reductive, that they rely only on the ignorance of the reader to be believed.

On the one hand, we need the hype to keep the money coming towards cancer research, but on the other, the hype not only messes with expectations and emotions, it negatively impacts on the work being done to treat cancer both now and in the future.

Misguided mainstream sentiment around cancer is dangerous.

We Don't Know What We Don't Know

In the UK, over half of adults under sixty-five right now will be diagnosed with cancer at some point in their lifetime.[5] It's the second leading cause of death globally, after heart disease.[6] Almost everyone will know someone who has, or has had, cancer in some form. And this global prevalence as well as personal attachment can sometimes make us forget how little we know about what cancer really is.

We all know it can be horrific, we don't all know exactly why it still remains a threat. And sometimes not knowing what we don't know is what stops us from seeing through the hype.

What Is Cancer?

First of all, cancer is not one disease. It's the name given to a collection of related diseases, which all come about when abnormal or damaged cells defy the body's usual elimination process and instead divide and spread, sometimes forming growths, which we call tumours.

It's a genetic disease, meaning the abnormal cells become abnormal due to some damage to the DNA within. The damage causes changes in the DNA, and these changes are called mutations. The DNA damage can happen for many reasons. Cells divide over and over throughout your life, and sometimes that division results in errors in a new cell's DNA. This is why cancer is more common in older people; it's simply a numbers game with more cell division having happened over the years. This is also why skin cancer is so prevalent; there's lots of turnover of skin cells, they divide far more frequently than other cells in other parts of the body. The damage could also be as a result of certain kinds of inherited mutations in our DNA from our parents. Or it could be as a result of environmental exposure: too much sun, too much tobacco or too much radiation, for example.

The genetic changes that tend to contribute to cancer normally affect three kinds of genes. There are the proto-oncogenes and the tumour-suppressor genes, which are involved in controlling cell growth and division, and when they are altered through DNA damage they allow the cells to grow and survive in an uncontrolled manner. There are also the DNA repair genes, which when damaged fail to fix abnormal cells.

This means that cancer cells are unlike normal cells in that they don't have the mechanisms within them to stop dividing when they become problematic, a process called 'programmed cell death' or 'apoptosis'. This is what leads to cancer spreading, tumours forming and, ultimately, the breakdown of processes and functions of the human body.

There are certain DNA mutations that are commonly found in particular kinds of cancers. But every cancer results from a combination of different genetic changes and, as cancer develops, more changes can happen as the cells continue to divide, meaning the abnormal cells within the tumour can differ from each other. That's all to say that there are many different kinds of cancers that can occur, for many different reasons, that all look different in each individual's body.

In short, each person's cancer is different and we don't always know exactly which changes in which cells have caused it, which is part of the reason it's so hard to treat.

WHAT MAKES CANCER EVEN MORE COMPLICATED

Some cancers result in a tumour, a solid mass of tissue. That tissue can be malignant, meaning it spreads into and invades nearby healthy tissues; others are benign, meaning they don't spread. Benign tumours, when removed, don't normally grow back; malignant tumours sometimes do. Benign tumours can normally be cut out with surgery, which is not normally life threatening unless you have one in the brain – the growth itself can do enough damage to kill you, never mind the spreading into and invasion of other cells beyond it.

Sometimes the cancer spreads beyond its immediate surroundings and into other parts of the body through a process called metastasis. If you have breast cancer and a metastatic tumour is formed on the kidney, you'd have metastatic breast cancer, not kidney cancer, because the metastatic cancer cells normally look like the cells of the original cancer. Most people who die of cancer die of metastatic disease.[7]

Cancer cells often hide from the immune system, meaning the body's internal defence system is rendered useless in flushing out the problematic cells. They can also sometimes influence the microenvironment around the tumour, resulting in blood vessels supplying them with oxygen and nutrients to continue to grow.

Making cancer even harder to treat is the variance among similar types of cancer. For instance, breast cancer is a collective of eleven genetically distinct diseases, each with different treatment options and prognoses.[8] And two people who have a similar form of cancer are likely to have different combinations of biological make-up, current minor or major ailments, ages and medical histories, meaning they are likely to respond to the same treatments in different ways.

There are so many options for type of cancer, expression of cancer in the body and response to treatment that it is extremely difficult to work out exactly what is going to work, quickly enough, for each individual.

Each research area focuses on a certain cancer situation, and due to so much variance in how cancer exists in humans, there are exponentially many areas for scientists to choose from. Funding for cancer research is split up among many efforts but, as you might imagine, the budgets shrink as scientists home in on specific areas.

TRYING TO GET RID OF CANCER

When there's a solid tumour contained in one area, surgery is normally used to remove the problematic cells, and sometimes it's the only treatment needed if the cancer cells haven't spread. Most of the time, though, surgery is paired with other forms of treatment.

There's radiotherapy, which uses high doses of radiation to kill off the cancer cells by damaging their DNA beyond repair so that they die off, break down and are removed by the body. Radiotherapy is normally administered over days or weeks, and then the cancer cells might still take weeks or months to die off after the therapy ends. Radiotherapy can be either 'external beam radiation therapy', coming from a machine that aims the beams of radiation at whichever part of your body has the cancer cells, or 'internal radiation therapy' where a solid or liquid source of radiation is put inside your body, travelling to the cancerous area.

Then there's chemotherapy, which uses drugs to kill the cancer cells by stopping or slowing their growth. Sometimes chemotherapy is administered in the form of pills, sometimes it is injected, sometimes it comes via an intravenous drip, sometimes it's a cream. Chemotherapy is normally given in cycles, with part of that cycle being the treatment and part of it being a rest phase so a patient's body can recover.

Both radiotherapy and chemotherapy target fast-growing cells in particular, which is what cancer cells tend to be, but we also have fast-growing cells that line our mouths and intestines, and that cause our hair to grow. The therapies are indiscriminate to some extent; their ability to destroy the cancer cells also applies to normal cells, meaning healthy cells can die off, causing hugely problematic side effects such as hair loss, nausea, fatigue and mouth sores. Adding salt to the wound, the side effects don't necessarily indicate how effective the treatment is being; sometimes you're left suffering from the side effects and the cancer is still thriving.

It's no secret that surgery, chemotherapy and radiotherapy can be trauma-inducing forms of cancer therapy, despite their relative success at getting rid of cancer cells. The side effects of the treatment are sometimes more difficult to live with than the cancer itself.

We all know that progress therefore needs to be made. And there are indeed many efforts pointed towards finding treatments that work when these therapies have been rendered ineffective, and alternative treatments that don't have such a damaging effect on the healthy cells in the body in the first instance.

It's here, in the dissemination of what's new and next in cancer treatment, where hype and misrepresentation of research start to happen. With so much depending on these new treatments working and becoming available, and therefore so much opportunity for those creating and selling them, it's an area ripe for problematic narratives and discarded nuance.

UNDERSTANDING THE GENETIC QUIRKS OF CANCER

The line of thinking that surgery, radiotherapy and chemotherapy follows is to remove or destroy as many of the cancer cells as possible in their entirety, while minimising collateral damage, and hope that nothing problematic is left over or grows back. Newer forms of therapy follow a different line of thinking: understand the characteristics or behaviour of the cancer cells and target the treatment specifically, in turn, reducing collateral damage and hopefully rendering the problematic cells useless in the process.

When researchers started to agree that cancer was indeed a genetic disease, the field started to focus its attention on what specifically was different in the cancer cells' DNA and, as such, how a treatment might be more effective if that understanding was at the core of its invention.

THE TARGETED THERAPY REVOLUTION

It began in the 1960s, when Dr Peter Nowell was investigating a rare form of blood cancer called chronic myelogenous leukaemia (CML). It was known to be a horrific disease; once patients were diagnosed, they rarely lived more than a few years and the treatments, which really only marginally extended life, were harsh and invasive.[9]

Nowell paired up with another researcher, Dr David Hungerford, and they went about analysing the cells of patients with CML. They found that one of the chromosomes in those patients, later called the Philadelphia chromosome after the city in which this was discovered, was slightly shorter than in everyone else.

A decade later, research continued on this peculiar Philadelphia chromosome, and Dr Nora Heisterkamp and her team worked out why the abnormal length occurred. She found that it happens when there are two normal chromosomes, one with a large chunk breaking off and one with a small chunk breaking off. Instead of their own broken pieces reattaching where they left off, the broken pieces swap over. The chromosome missing the large piece now has the smaller piece attached, making it smaller overall. This hybrid gene is the Philadelphia chromosome, and it was known as BCR-ABL as it formed from the fusion of the broken pieces BCR and ABL.

In 1980s Los Angeles, Dr Owen Witte and his team proved that when BCR-ABL forms in blood cells, it causes CML.

In the 1990s, Dr Brian Druker and Dr Nicholas Lydon at Oregon Health and Science University wondered if a drug that blocks the activity of BCR-ABL might then be able to treat CML. In 1998, they kicked off a clinical trial testing their chosen drug STI-571, later called imatinib, and five years after the trial, 98 per cent of the patients were still in remission.

The US Food and Drug Administration (FDA) approved imatinib as a treatment for patients with CML who have the Philadelphia chromosome in 2001. Nowadays, someone who has had two years of imatinib treatment and is in remission has the same life expectancy as someone without cancer.[10]

Imatinib was not only a hugely important discovery for those with CML. It also kicked off the targeted therapy revolution, and gave researchers confidence that using unique genetic understanding and patient-specific approaches could work in treating cancer the world over.[11]

FINDING THE TARGETS

Since 2001, more targeted therapies have been approved for usage world-wide, and some have cemented themselves as standard treatments for many different forms of cancers. Researchers are learning more about what exact changes are happening in the cells that lead to cancer, which, in turn, helps them design therapies to specifically target that change.

Targeted therapies tend to be either small-molecule drugs, which enter the cells themselves and focus on target processes inside, or thera-peutic antibodies (also known as monoclonal antibodies), which are proteins that attach to the outside of the cancer cells and do their work from there. The specific jobs these targeted therapies are tasked with might include stopping the cancer cells from growing, or stopping the signals that help form the blood vessels that feed the tumour, meaning it 'starves' the cells to death. They might be tasked with attracting cell-kill-ing substances towards the cells, almost like little flags on their surface for them to spot. Hormone therapy is a form of targeted therapy for some types of breast and prostate cancer, and these therapies might prevent the body from making the hormones that help the cancer grow, or halt the hormones from acting on the cells.

A specific form of targeted therapy focuses on the immune system – either boosting it so it can work better against the cancer, or simply flag-ging the cells so they can no longer hide from it. This area of cancer therapeutics is called immunotherapy, and is one of the most hotly discussed and theoretically promising areas of research today.

THE BODY'S OWN LINE OF DEFENCE

The idea behind immunotherapy is pretty simple: prompt the body into using its own line of defence to get rid of the problematic cells created in the body in the first place. Actually making that happen effectively is, unfortunately, easier said than done.

Immunotherapy is a biological therapy, meaning living organisms are used to create the substances in the treatment, and there are various different kinds. There's BCG (Bacillus Calmette–Guérin), a weakened form of bacteria that causes tuberculosis, which can prompt a stronger immune response for bladder cancer. Or there's the checkpoint inhibitors, which block the proteins that prevent the immune system from responding to cancer cells. There are the therapeutic antibodies, which specifically flag the cancer cells to the immune system by attaching to the outside and essentially putting out a call for help. And then there is adoptive cell transfer (ACT), which boosts the natural ability of a body's T cells (a type of white blood cell found in the immune system) to attack the cancer cells. For ACT, T cells are taken from a patient's blood, and the most effective ones for fighting against the cancer are grown in huge batches in the lab over the course of a few weeks, and then reinserted back into the patient's body.

In 2011, the first immunotherapy was approved by the FDA: a checkpoint inhibitor called ipilimumab (marketed under the marginally easier to say name 'Yervoy'). At the time of writing, there are almost 3000 clinical trials, testing all different kinds of immunotherapies, registered worldwide.[12]

Immunotherapy has become a very hot area of cancer research and, as such, a regular feature in the hype surrounding the field. When the therapy works, the results are terrific, but so few cancers are eligible and thus tiny proportions of those with cancer have access to these treatments. A study in 2017 showed that almost 70 per cent of Americans predicted to die of cancer would die of a type of cancer, including colon, breast and ovarian cancer, for which there is no FDA-approved immunotherapy option.[13] Sometimes the tumours have morphed by the time the T cells homing in on them arrive, or the tumours find entirely new ways to grow despite both the treatment and other supporting drugs. Relapses have also been found to happen even after the therapy is complete, with patients having gained little if any more time before dying.[14] Side effects can also be harsh, so the quest for a better experience of therapy versus chemo- or radiotherapy hasn't quite been fulfilled.

Many new therapies, particularly for advanced diseases such as terminal forms of cancer, aren't perfect. But the hype around immunotherapies runs the risk of severely damaging patient expectations, trust and hope.

'IT MUST BE TRUE, I SAW IT ON TV!'

In the US, the Super Bowl is known for drawing one of the biggest audiences in the annual TV line-up. And because so many eyeballs are simultaneously glued to the screens, the ad breaks are packed with commercials costing the companies that run them about $5 million each.[15]

In 2016, pharmaceutical company Bristol-Myers Squibb ran a Super Bowl ad for its immunotherapy called Opdivo.[16] The ad offered 'a chance to live longer' with Opdivo, claiming that its immunotherapy treatment for certain forms of lung cancer (to be prescribed only after conventional chemotherapy has failed the patient) 'significantly increased the chance of living longer versus chemotherapy'.

At least, that's what the voiceover and large capitalised letters on screen claimed. The small print read: 'In a clinical trial, Opdivo reduced the risk of dying by 41% compared to chemotherapy. Half the Opdivo patients were alive at 9.2 months versus 6 months for chemotherapy.' The ad ended with the voiceover: 'A chance to live longer. Ask your doctor if Opdivo is right for you.'

A 'significant increase', arguably, is subjective, but personally I'm not sure three extra months on a drug with such severe side effects that it was reported that some patients had to stop taking the drug is really worth it.[17] 'Significant increase' also, without reading the small print, means something entirely different to scientists and industry analysts as opposed to patients; 'progress' might mean 'cure' to a hopeful patient.[18]

Opdivo is priced at around $150,000 for the initial treatment plus $14,000 per month, and not all insurers would cover that cost. In the same year as the Super Bowl ad, the National Institute for Health and Care Excellence (NICE), the government body in charge of deciding whether a drug is worth offering through England and Wales's National Health Services, rejected Opdivo as too expensive in terms of the benefit it was meant to offer.[19]

Direct-to-consumer advertising for prescription drugs is illegal in most countries; only in the US and New Zealand are people subjected to billions of dollars' worth of ads showing surprisingly healthy-looking actor-patients flogging poorly described therapies for people who are dying.

And this kind of advertising is only increasing in the US, with Kantar Media reporting that drug companies spent over $6 billion on direct-to-consumer advertising in 2017, up 64 per cent since 2012.[20] Not only is the spending shifting, the content of the ads too has progressed. A study comparing drug ads from 2016 versus 2004 found that the emotional framing of drugs as helping people gain control and/or social approval had gone up, while the factual information, biological explanations, and discussions of causes, prevalence or risk factors had all decreased.[21] Research also shows that despite doctors broadly being against this kind of advertising, the marketing does contribute to doctors prescribing more of these drugs.[22]

To a British person watching an ad for a drug, the way the risks and side effects are read out may be almost comical. But it's also worrying to hear the list monotonously read out over a video of families living their best life, while also considering the mindset of patients struggling with a late-stage disease, who would most likely watch the ad with hope, possibly miss the small print, and tune out the list of side effects as if it was simply a list of stock terms and conditions.

Immunotherapy is an incredible advance in the quest to treat more cancers better, but communication around its efficacy, availability and risks must be clear if the trust and hope of patients, and those who care about patient welfare, are not to be meddled with and lost.

A SPECIAL IMMUNOTHERAPY

There's a kind of immunotherapy called CAR-T therapy, which makes the most headlines. It's a type of ACT, where instead of just seeking out the most effective T cells and growing them in large numbers, the T cells are modified in the lab to make them better recognise the cancer cells and make them more deadly in killing them, before then being reinserted into the body. (The modification includes producing special structures on the surface of the T cells called chimeric antigen receptors [CARs], which is what enables them to recognise the proteins on the cancer cells, hence the name of the therapy.)

CAR-T makes headlines for two main reasons.

CAR-T WORKS REALLY WELL (SOMETIMES IN SPECIFIC CASES)

First, CAR-T therapy has been proven to work for some patients for whom all other treatments haven't worked. The trials were small and the forms of cancer are very rare but, nonetheless, patients who were very close to death who had exhausted all other options – including in one instance, a one-year-old called Layla – went into remission after their CAR-T therapy.[23]

In 2017, the FDA approved the first two CAR-T therapies. Novartis had its Kymriah therapy approved for the blood cancer B-cell acute lymphoblastic leukaemia (ALL), and Gilead (only one week after acquiring the biotechnology company Kite Pharma, who created it) had its Yescarta therapy approved for another blood cancer called aggressive B-cell non-Hodgkin's lymphoma.

Amidst the patient success stories are those not so fortunate. In March 2017, Juno Therapeutics had to end its lead CAR-T programme after five patients died from brain swelling caused by the therapy, and in the trial that paved the way for Gilead's Yescarta to be approved, three deaths were reported as a result of side effects.[24]

One of the side effects that can make CAR-T deadly is a response called cytokine-release syndrome (CRS). The release of cytokines is a good thing, as their presence demonstrates that the T cells are at work, but if too many cytokines release into the bloodstream, huge drops in blood pressure and high fevers can be fatal. The good thing about CAR-T, though, is that it's designed to be a one-time therapy, so once the T cells have done their work, the side effects should subside and the therapy shouldn't be needed again.

CAR-T has so far also not shown as much progress against solid tumours such as in breast or lung cancer. There have also been reports of patients going into remission after only a couple of years.[25] And then there are issues with the cost of the therapies: in the US, Kymriah is priced at \$475,000 and Yescarta at \$373,000.[26] When you factor in the hospital costs and everything else surrounding the procedure, we're talking about \$1.5 million per patient.[27] Scalability, and therefore reduction in cost, of the therapy still needs more investigation, as well as a reduction in time from T-cell extraction to reinsertion to ensure the patient doesn't die while waiting. Researchers are therefore testing ways of using immune cells from healthy donors to create 'off the shelf' therapies, which would be immediately available for use, but these are still at a very early stage.[28]

In short, CAR-T is showing potential as an option for people with very specific advanced cancers, who have exhausted all other options. If this is to be made mainstream, much more research still needs to be done. Hype surrounding positive patient stories is understandable, but whenever CAR-T is described as the 'miracle cure' for cancer, it must be put into the context of where the broader science is currently at – and right now, it is simply not a 'miracle cure'.

CAR-T Plus CRISPR Equals Hype

The second reason CAR-T gets so much attention is because of its link to genetic modification. CRISPR (clustered regularly interspaced short palindromic repeats), a relatively recent breakthrough that allows for simple, fast and cheap editing of DNA, happens to be a pretty trendy topic at the same time CAR-T is emerging, and when it is paired with the 'cure for cancer' narrative, the headlines almost write themselves.

There's reason to be very excited about CRISPR. Genetic editing without it can be a costly, inaccurate and time-consuming process, so this cheap, precise and quick-to-use tool looks like it could be the democratiser of the gene-modifying space. That has repercussions not just in cancer treatment, but in many other areas of health, as well as agriculture. In 2011, there were fewer than 100 published papers on CRISPR; in 2018, there were over 17,000.[29]

CRISPR, in theory, should be able to speed up and bring the cost down of CAR-T therapy due to its shortening of the gene modification process from weeks to days. The problem with CRISPR, though, is that it isn't perfect. In 2018, it was found that its approach to gene editing can delete and rearrange other parts of DNA not meant to be modified, and could ultimately kill a person should those genes – edited both purposefully or accidentally – find themselves back in a human body.[30] Other studies in 2018 found that CRISPR-edited genes could actually trigger cancer further on down the line.[32]

Without the knowledge of how CAR-T and CRISPR work, and without the challenges and risks associated with them being fully and clearly conveyed alongside the success stories, it's understandable why many people are excited. The reality is that while CAR-T is brilliant for very specific cases, until it proves to be more transferable to other forms of cancers, we cannot afford to pin our hopes and expectations on it without knowing the whole story.

THE RISE OF PERSONALISED TREATMENT

Deciding which treatment is best for each individual patient has been, until more recent years, a task that has relied on the experience of medical professionals, the availability of new research to reference, and a bit of luck. After tests and scans and biopsies of bits of the tumour, the doctors would assess what was most likely to be happening within the cancer cells, compare the patient to others who have been at a similar stage with similar types of cancer, and then match the cancer to treatment that has tended to work in the past. This approach isn't now relegated to the history books; much of the investigation into cancer still works exactly like this. The difference nowadays, though, is that there's increasing availability of a powerful tool that allows us to peek inside the cancer cells of each individual, and see for sure what's really going on.

Genetic testing allows us to read DNA. When we can read DNA, sometimes we can translate that code into understanding what's going on inside a living thing. Each string of DNA represents production of certain proteins that, in turn, lead to certain behaviours of cells. If we can understand the role of every single string of DNA, when we manage to read out the DNA, we should be able to know exactly what is going on in each and every cell.

This becomes extremely useful in understanding cancer, as the DNA in those cells differs from your regular DNA, so if you take a sample of the cancer cells and compare their DNA to your baseline version, the exact functionality of those cells can be uncovered and then matched up with a suitable treatment to target their specific behaviour.

The advent of cheaper genetic testing, along with more research going into understanding what the strings of DNA translate to, means that there is much excitement and anticipation around the potential of genetic testing in bringing a personalised approach to cancer treatment.

An example might be a patient with some form of lung cancer, having their tumour sequenced to try to spot a mutation in the gene called EGFR, which provides instructions for making a protein called the 'epidermal growth factor receptor'. Certain mutations, which then cause the cells to divide rapidly, are found in some kinds of lung cancer, so targeted therapies that might inhibit EGFR could then be recommended for that specific patient. Tumour sequencing can lead to finding specific mutations that directly match available treatments, as opposed to the patient's cancer being segmented by geographical location on the body or by stage and then matched with what has been used in the past for those types of cancer, without the deeper understanding of their unique case. Another benefit to having the tumour sequenced is that the results can sometimes predict whether a particular treatment option is going to work, so that the patient can have more of an idea of efficacy before embarking on a particular route, with all the likely side effects taken into account.

Indeed, the potential of the technology is huge in saving lives, making quality of life during treatment higher, and therefore saving healthcare costs for the patient (or payer) and the system. Genetic sequencing could quite easily come across as the answer to many of the problems inherent in cancer therapy.

THE UNFORTUNATE REALITY

Sequencing cancerous tumours is a brilliant innovation but there are flaws, which rarely get coverage, meaning it is often presented without crucial nuance.

For starters, the cells that make up the cancer as a whole tend to be heterogeneous, meaning they aren't all the same. If you take a sample of a tumour and sequence it, you only get information about the cells you investigate. A biopsy might not accurately represent the cancer as a whole and thus any treatment recommendations made off the back of the information have to take into account the limited nature of the findings.

Not only is a tumour heterogeneous, it changes over time. Treatments that may be indicated to work by one test may now be facing an evolved tumour. Tumours also can become resistant to therapies over time. Sequencing can only capture a snapshot of sorts so, for it to be effective, multiple rounds of sequencing would have to take place, bringing up the overall cost, which patients may not be able to afford, insurance companies may not cover or national health services may not be able to justify.

Next, the irony of sequencing cancer cells in order to take a more personalised approach to treating them is that you need the right kinds of cancers for it to be really worth the sequencing. Only certain genetic mutations have matching therapies to counter them, meaning only certain kinds of cancer will even warrant a genomic test in the first place. For example, people with broad cancer types such as melanoma and leukaemia, and some forms of breast, colon, lung and rectal cancers, are likely to have their cancers tested at the point of diagnosis to see if there's either an existing therapy or a clinical trial that may be relevant for them. If you have a cancer with little or no targeted therapies or trials available (or the doctor simply doesn't know about them), the DNA tests are unlikely to be offered.[33] In 2018, a study showed that of all the people who have metastatic cancer in the US as of 2018, only 9 per cent of them will have a mutation that has a matching FDA-approved treatment available.[34]

The specificity of the testing isn't just a matter of that lack of treatments that match, but a lack of understanding of what the actual DNA readout means in terms of the corresponding behaviour of the cancer cells. When a specific test is prescribed, only certain parts of the DNA are looked at; the bits that researchers already know have corresponding knowledge about cancer cell behaviour attached to them. If researchers are to look at the full DNA code of a cancer cell, and find differences between that and the corresponding normal cell, those differences might either already be understood to not cause cancer, or might not yet mean anything at all to researchers; the science simply might not have been done in analysing that part of the code. Not only does this not correspond to a treatment option, it doesn't even correspond to an understanding of the cell itself. How demoralising it must be to be told categorically what the difference is in the DNA of your cancer cells, but also then be told that no one knows what to do about it.

Genetic sequencing hasn't been around long enough yet for all the genetic changes that cause cancer to be discovered. There are so many variables that go into how a human responds to treatment that, even in another decade when we're sure to have a far better understanding of the genetics of cancer cells, we won't understand the full nature of a person's cancer or be able to predict their reaction to certain treatments. There's the person's age to take into account, as well as any other ailments they may have and their medical history.

With such uncertainty associated with cancer, particularly with the forms that have had less of a research focus over the years, there might not be medical consensus around what route is best for every patient. Of course, doctors will do their best to explain different treatment options to patients, but when second or third opinions differ, the process of deciding your own fate can be not only highly confusing but traumatic in and of itself.

Headlines proclaiming miracle cures, cancer breakthroughs and medical panaceas mislead us into thinking we're not far off 'beating cancer', and the realisation that this isn't the case, while you're in the doctor's surgery talking about your own cancer, is not only heart-breaking, but avoidable.

HYPE IS A DOUBLE-EDGED SWORD

While the bulk of this chapter so far has focused on the overhyped narratives and the issues of getting patients' hopes up with bombastic headlines and the promise of a soon-to-be found 'cure', there's another side to the cancer hype coin.

As the cancer research field opens up with more scientists focusing on more approaches to treating cancer, more money is needed not only to conduct the research in the first place, but also to spin the findings out of the lab and into companies, and to take the drugs to market. Positive public sentiment is therefore required to continue putting pressure on governments to fund cancer research, to prompt charitable giving, and to create a feeling of excitement and opportunity around startups popping up in the cancer therapeutics space. The hype prompts investors to fund new biotechnology startups or bigger drug companies to buy these companies and then offer more new treatments at scale.

In short, whether the hype is warranted or not for specific headlines, general excitement around what's next in the cancer-therapy space contributes to more interest, more money, more available drugs and more campaigning, which, it would seem, can only be a good thing.

Misleading headlines don't just contribute to misunderstanding, though. The hype isn't problematic just because some people 'don't get it'. The hype keeps us backing what's currently going on; the investment, the focus and the societal support. But what if what's currently going on isn't always best for moving cancer therapeutics forward? What if the research we are cheerleading, without knowing the ins and outs of it, isn't actually the best possible research? What if, in our deep desire for cancer therapy to advance and for patients to have better options, we miss the problematic parts of the broader field, and 'accidentally' cheer on misguided research without realising it?

Rooted in much of the hype around advances in cancer therapeutics is an accidental disregard of bigger issues at play.

We don't know what we don't know, and *that* is damaging cancer research.

OF MICE AND MEN

At the centre of biomedical research is the humble mouse. They are anatomically, physiologically and genetically similar to humans, as well as being small and with a short life span, and thus are regularly chosen for testing new treatments coming to market.

But just because a drug seems to work in a mouse doesn't mean it'll work in a human. And just because a drug doesn't kill a mouse doesn't mean it won't kill a human.

Because testing in mice isn't enough of a check before giving out new treatments to patients, multiple rounds of clinical trials take place afterwards to ensure not only that the new treatments aren't toxic to humans, but that they have some kind of demonstrable benefit versus what's already out there. The difference between a study conducted on mice and one on humans can be about five years and millions, if not billions, of dollars.[35]

A 2017 study found that words such as 'revolutionary', 'life saver', 'breakthrough', 'marvel' and 'game changer' are regularly associated with studies done only on mice. In fact, 14 per cent of the articles they analysed described new drugs with these superlative terms, which had only been proven to work on mice.[36] That means 14 per cent of those articles were proclaiming huge advancements in breast or colon or lung or blood cancers, most likely being read by people who currently have those diseases, without any human testing whatsoever.

It's become a bit of an industry inside joke: 'I've got a whole lab of mice I can cure cancer in.' But that sentiment is being lost in translation.

The finding from that same study that was more problematic, however, was this: the researchers found that those same superlatives were being published in news stories about drugs that haven't yet been approved by the FDA. Not 14 per cent of the time, not 30 per cent of the time, but 50 per cent of the time.[37]

Half of those stories were referring to science that hadn't yet been allowed to leave the lab. The science may have been shown to be promising, but there was no guarantee that the treatment would ever reach the market.

Of course, many people won't even read or register the clarification, if there is one, that the drugs haven't been tested on humans or been approved by the FDA. Most people will read the headline and assume that means the drug is available now.

It's a relatively common understanding that discovering new drugs is not a quick task, but leading those on the outside to believe we're 'there' with these kinds of superlative headlines, when there's still so far to go, reflects a lack of understanding both by those who are writing these pieces and those reading them about how drug regulation and approval works.

Over-simplification of trial results is problematic. Beyond not clarifying the time to market, the type of study or if the drug has even been approved, the superlative language acts as a dangerous shortcut. The person reading the article might assume that, because what they're reading about has been described as a 'breakthrough' or even 'cure', they don't need to doublecheck the source. The trust in the publication, in those cases, blocks critical thinking.

There's something to be said here about the inaccessibility of scientific papers and clinical trial reports. Sometimes the trial reports aren't fully published as they are seen as proprietary by the pharmaceutical company. Sometimes the papers are behind the paywall of the scientific journals in which they're published. And when you actually can access the reports, they're mostly written in language only decipherable by academics in the immediate field surrounding the research, and are therefore quite impenetrable for everyone else to read. Science journalists and media outlets, therefore, have the task of translation.

When the Drugs *Do* Get Approved

When the translation is done in such a way that superlatives are misused, misunderstanding takes place. And that misunderstanding is problematic when those on the outside not only don't know to check whether the approval has happened or what kind of trial has been completed, but also don't know *how* to check how successful the approved drugs really are.

There's much discussion in the pharmaceutical industry on what can be deemed a fair and useful metric for the success of a drug. If a company wants to prove that the drug gives patients an extra ten years of life, then the only sure-fire way of proving that would be to run a trial with many patients for over ten years and monitor morbidity throughout. This is not only costly, it extends the time-to-market for potentially life-saving drugs. If the regulators required pharmaceutical companies to do these kind of trials for every new treatment, we'd be even further from getting good new science into the hands of the people who need it most.

Instead, when clinical trials are run, there are many other ways of measuring success to extrapolate the worth of the treatment, so the FDA can decide if it should be on the market. These substitute measures are called 'surrogate markers' and they include 'progression-free survival', the length of time the patient lives with the disease without the cancer getting worse, or 'tumour shrinkage'. These markers sometimes correlate with a real clinical endpoint, in other words, overall survival gain, but not always. If the FDA approves a drug based on surrogate markers to get it to patients faster, the idea is that a follow-up study will test for the clinical endpoints and hopefully reinforce the early approval.

It's key to point out here that when the FDA (or another country's drug-approving body) gives its blessing for a drug to be on the market, the shares of the company behind it are likely to go up. There's no 'traffic light' system for approval – the drug is either in or out – meaning that an FDA approval is the point at which a company can set a price and start selling its treatment, regardless of how good it is. If nothing else has been approved that is similar to said drug, of course, the prices can skyrocket. An early approval counts just as much as a regular one.

So if it's not clear from the reporting of these trials (or the scientific papers themselves) how effective the drug actually is, drugs with only an incremental gain in terms of survivability, with possibly huge negative side effects in tow, might make it onto the market with much fanfare. And when that happens, sometimes the associated critical analysis around whether it is really worth the cost (and side effects) is missed out.

In October 2018, Dr Peter Schmid of St Bartholomew's Hospital in London caught the attention of the cancer research world when he presented his team's findings at the European Society for Medical Oncology Congress in Munich, simultaneously publishing the results in the *New England Journal of Medicine*.[38]

The new treatment he and his team had been testing was a combination treatment, an immunotherapy drug called atezolizumab (marketed by Roche as Tecentriq) paired with a previously approved chemotherapy drug branded as Abraxane. The team had run a trial on 902 women, randomly assigning them to one of two groups: 451 patients with chemo plus placebo, the other 451 patients with chemo plus atezolizumab.

The reason there was excitement over this trial was the type of cancer in question: triple negative breast cancer (TNBC), a particularly aggressive form. It's called 'triple negative' breast cancer because it means the cells don't have any of the three receptors that can respond to the hormone therapies or other drugs on the market: oestrogen, progesterone or the protein HER2.[39] As such, patients with TNBC have a median overall survival rate of less than eighteen months.[40] Immunotherapy, until October 2018, wasn't proving much use for breast cancer, so research focusing on this area was of great interest to the field.

Schmid spoke about the results of the study, saying that the researchers had seen 'a nearly 10-month overall survival difference, which I think is a key encouragement to see this drug as a new standard'.

The ten-month overall survival difference, though, was a considerable overstatement. It was based on surrogate markers.

The study split the patients up into the placebo group and the treatment group, and also split the patients into people who had a particular protein called PD-L1 and those who didn't. The results showed progression-free survival (time alive, with cancer, before it then progresses again) to be statistically significantly longer in the patients treated with atezolizumab versus chemotherapy alone. They saw this both in the entire population, with patients' cancer not progressing for 7.2 months, up from 5.5 months with chemo alone, and also in the PD-L1-positive patients, with a 7.5-month progression-free survival, up from 5 months with chemo alone. Some of the PD-L1 reached ten months progression-free survival (which is where Dr Schmid's claim came from), but not enough of these patients registered this length of time for it to be statistically significant.

Overall survival wasn't formally assessed, so with a two-month progression-free survival rate and unknown overall survival benefit, it seemed a bit premature to be calling it a 'new standard of care'. It's also worth mentioning there was an increase in side effects, nothing too unexpected, but toxic and very unpleasant nonetheless. For two months' extra living, versus chemotherapy, with the cancer still there and high toxicity included, there are therefore questions around the worth of the treatment versus solely using the chemotherapy drug already on the market.[41]

Not long after the announcement, the FDA announced an accelerated approval process to get atezolizumab through the system more quickly for PD-L1-positive patients, based on these results. They granted a Priority Review of six months, as opposed to the standard ten months, and in March 2019 the FDA approved the drug for use in this specific subgroup.[42]

Around the time of the initial research publication, Roche presented its expected revenue opportunity for Tecentriq (its version of atezolizumab) to be between $500 million and $1 billion.[43] Market analysts estimated Roche had a lead of about eighteen months against their biggest immunotherapy rivals, and with Merck's similar trial failing six months later, Roche had the only drug of this kind on the market when the FDA gave its blessing.[44]

BAD INCENTIVES, NOT BAD PEOPLE?

Tecentriq is one of many drugs, priced at over $100,000 per year, that has won FDA approval based on proving benefits on surrogate markets but not proving survival extension.[45] Numerous trials are designed to measure many of these markers in the hope that one of them will show some kind of significant result to justify FDA approval.

As indicated earlier, when the FDA approves such drugs, it tends to also stipulate that continued approval may require or be contingent on confirmatory trial data. In other words, it says that for the drug to continue to be on the market, the ultimate clinical markers such as overall survival rate will have to be proven.

Only 14 per cent of drugs approved on the basis of surrogate markers, however, later prove to have a survival benefit after about four years on the US market.[46] And not all approvals based on surrogates require this extra study, meaning some drugs, which initially had accelerated approvals to allow them to be sold, are never actually confirmed to extend life overall; not all surrogates are known to be correlated to overall survival. The FDA's own recommendations state that approvals must be based on the established surrogates – the ones known to have a relation with overall survival – but it doesn't always demand that these validated surrogates be used. Indeed, for 37 per cent of regular approvals, there is no validation in the entire medical literature.[47] The same 2016 study also proved that in only 10 per cent of approvals is there a strong, proven correlation between the surrogate and survival. It seems, then, that the FDA is failing to follow its own recommendations.[48]

Drugs that, ultimately, cause harm that isn't first observed in the initial study might therefore also be missed.

Cancer drugs are hugely profitable these days, with recently approved drugs bringing in revenues of around $1.7 billion after only four years on the market.[49] Considering the fact that many approved cancer drugs carry with them around fourteen years of exclusivity for the companies that own them, this is clearly a conservative estimate of the overall benefit for a pharmaceutical company's shareholders of having such a drug on the market.

This loophole of sorts encourages the pharmaceutical industry to push forward on these kind of interim surrogate marker trials and take to market expensive drugs that arguably aren't worth the cost. Of course, for many people, an extra two months on Earth are worth the side effects and the cost, but incentivising this behaviour might stifle innovation in other areas if pharmaceutical companies are being rewarded monetarily for developing drugs that extend life by such a short period. The harder science necessary to extend life comes at a huge research and development cost; if pharmaceutical companies are satisfying their shareholder demands on drugs with little to no effect, where's the incentive to push them to work on better science?

In fact, when a 2018 study looked into this pursuit of marginal benefits and a so-called 'me-too' mentality stifling innovation and creativity, the analysis found that of seventy-one drugs for solid tumours, the average survival benefit was only 2.1 months.[50] Almost two-thirds of cancer drugs are being approved solely on the basis of surrogate markers, as opposed to improvements in quantity or quality of life.[51] Vinay Prasad, associate professor of medicine at Oregon Health & Science University, is a noted vocal critic of cancer research's direction, and wrote in one paper:

> Drug regulators' acceptance of any statistically significant improvement shown in a single randomised trial and lofty drug prices has created a situation where it is now, hypothetically, profitable for a company to run a clinical trials portfolio of chemically inert compounds ... We certainly do not believe that pharmaceutical companies are actively pursuing ineffective drugs, although the current oncology drug development and regulatory environment does little to discourage such an agenda.[52]

It's reductive to say that if we, as a society, stop buying into the hype around new drugs that these kind of problematic incentives and practices will go away. But there is an argument to be made that if more questions are asked about what's really being approved, we can open up more discussions about what 'significant benefit' and 'significant increase' really mean when it comes to cancer drugs. Of course, each of us will have different answers to that question based on our own experiences of cancer, but without that discussion happening openly in the public sphere, the FDA will continue to approve what many people feel are inadequate drugs and, in turn, continue to encourage the drug companies away from doing better.[53]

PUTTING SOCIETY FIRST?

When looking into the problems related to cancer therapeutics, it's impossible to not also consider the difference between public and private health systems.

Of course, in countries such as the US where citizens or their employers pay for health insurance to cover costs of hospitalisation and drugs, the debate around worth of drugs centres more on what the drug companies price them at and how insurance companies decide what to cover. In countries such as the UK, however, the discussion centres more on what is deemed a worthy investment considering the spread of healthcare funding across the nation's citizens. A drug that gives the patient only two months of extra life is deemed worthy in the US if the patient can find a way to pay for it and wants it; in the UK the decision is much more centred not on the individual but society as a whole.

The public health system can feel inhumane to some degree; it relies on considering the value of one life versus others to an extent. But the system does tend to be less wasteful in its spending on inefficient and arguably ineffective drugs.

In England and Wales, sometimes there is controversy surrounding what NICE approves and rejects, especially when it comes to making decisions about expensive drugs that add little time to dying patients' lives. NICE has to weigh NHS funding across all diseases, drugs and citizens.

So in the 2010 UK election campaign, in which the Conservative Party was challenging the incumbent Labour government for power, the Tory leader David Cameron promised to create a new fund so that people could get access to the latest treatments that NICE wasn't approving. Of course, this kind of narrative, fed to voters without the nuance around efficacy and cost of these new drugs, was a great success. And when David Cameron was elected to be the Prime Minister that year, the Cancer Drugs Fund came into being.

The fund ran from 2010 to 2016, ended up costing the UK taxpayers £1.27 billion, and was eventually deemed both a huge waste of money and a major policy error.[54] In 2015, the fund's costs had reached unsustainable heights, and more than half of the drugs that the fund had been set up to cover were delisted as a result. It was found that only 18 per cent of the drugs covered by the fund had met internationally recognised criteria for being deemed beneficial to patients, the average extra survival time across the drugs was only 3.2 months, and it was concluded that a majority of patients covered by the scheme may well have actually suffered more than were helped due to side effects.[55] The Cancer Drugs Fund was eventually consumed by NICE, which still comes under fire from the general public.

Whether to include an extremely expensive drug with limited or unproven efficacy on the list of treatments supplied by public funds is not an easy decision for a governmental body to make. If the drug ends up being proven to be hugely impactful further on down the line, the government body would be criticised for a delay in approving the treatment; if it's found to be damaging or ineffective, it will be criticised for not putting the money to good use elsewhere. Either way, a government-run scheme is at the mercy of public sentiment, especially when government ministers are often elected based on their claims and plans surrounding the NHS. The incorrect claim by Vote Leave that Brexit would save '£350 million a week' for the NHS was a powerful lie many people believed and voted accordingly.

Hype, therefore, plays an extremely important role in guiding public health funding decisions, whether the hype is warranted or not.[56] And patients can only look on while the pharmaceutical companies and the government bodies negotiate on price to get the drug into public healthcare systems. Without the nuance of efficacy, price, surrogate and clinical markers, and knowledge of side effects, the patients are often left frustrated and heart-broken when what's perceived as a viable treatment for their life-limiting disease is rejected or endlessly debated.

THE GOVERNMENTAL MORAL BACKBONE

The idea that, as an individual, you can have some level of influence on a huge, complex field such as cancer therapeutics can feel untrue. It can seem like an impenetrable area of research, which only those with lab coats or corporate pharmaceutical power can really have much of a say in.

The truth is, though, that government has much more control over cancer research, and healthcare as a whole, than we're really led to believe. Much of the narrative around problems in the system put the blame on pharmaceutical and insurance companies and, yes, they play a huge part in propagating and creating problems, but without the incentive structures, the allowances and the negotiation of central government, those companies cannot function or profit. Best-case outcomes in the industry rely, then, on the willingness of government to take action, fuelled by pressure from the public.

NOVARTIS VERSUS INDIA

Remember imatinib, the first targeted therapy to be proven to work and approved by the FDA? It turns out this is a wonderful example of a government being able to take control of the power of the drug for the benefit of the majority of the population, despite worrying threats that a powerful company was making. It's an example that shows that government, and by extension, people, can have the power when it comes to defining what makes most sense in the cancer therapeutics world.

Swiss pharmaceutical giant Novartis is the company behind Glivec, the branded version of imatinib. In 2005, Novartis decided to file a legal challenge against the Indian government, preventing any form of imatinib other than Glivec being sold in India, which kicked off a long battle pitting corporate interests against the health and wealth of an entire nation.

The reason that this blew up in 2005, and not at any other time in the history of India's pharmaceutical efforts, was due to the country joining the World Trade Organization (WTO) in 1995. Before then, the Indian government didn't allow patents for drugs, which led to India being known as the 'Pharmacy of the Developing World'. Not only did Indian companies provide to Indian patients, they provided large quantities of cheap drugs to many other developing countries. (Alongside China, Brazil and Russia, India at the time led a group of seventeen high-growth pharmaceuticals markets, referred to by the hilarious term 'pharmerging countries'.) India supplies drugs for 80 per cent of the 6 million people receiving HIV and AIDS treatment in the developing world, and it is the second leading provider of medicines distributed by UNICEF in developing countries.[57] India's production of HIV and AIDS medication had led to a dramatic lowering of costs from about $10,000 per year in 2000 to $150 in 2014.

When India joined the WTO in 1995, the government was given ten years to comply with an international intellectual property rule called the Trade-Related Aspects of Intellectual Property Rights (TRIPS) Agreement. So, in 2005, drugs were essentially allowed to be patented in India. TRIPS could be interpreted by each country with respect to its own laws and socioeconomic conditions, and, as such, India had a provision in its own Indian Patents Act preventing a practice called 'evergreening' – where pharmaceutical companies extend a patent on a drug by making slight changes to it.[58] Section 3(d) of this Act stated that patents wouldn't be awarded if new versions of drugs did not demonstrate any significant changes in therapeutic effectiveness over pre-existing forms, thus preventing TRIPS from upending India's booming pharmaceutical industry.

Now, when Novartis started working on Glivec, it filed patents in many countries in 1993 to protect it from other pharmaceutical companies, but it didn't include India as it did not have drug patent laws at the time. In 1997, when Novartis developed a newer, salt form of the drug, it filed for a new round of patents, now including India, thinking that by 2005 its government would have to honour the patent due to TRIPS.

By the mid-2000s, Novartis executives had grown impatient with India's lack of decision on the company's patent application, and put pressure on the government to do its review. Due to that pressure, the Indian government agreed to grant Novartis exclusive marketing rights for Glivec in the meantime, and the company quickly set about getting injunctions against Indian companies that had already been selling generic forms of imatinib since its FDA approval in 2001.

As a result, the price of imatinib in India shot up about twenty times, essentially overnight, and many public-interest suits were quickly filed petitioning for the patent application to be rejected and for the generic companies to get back to providing India's patients with the drug.

When Novartis's patent application was rejected in 2006 under section 3(d) of the Indian Patents Act, a five-year legal battle ensued.[59] Novartis decided not only to contest the rejection decision, but the Indian patent system itself, claiming section 3(d) didn't comply with TRIPS. It was clear in 2005 that if Novartis were to win its case, access to medicine in India across the board would be compromised, affecting all other drug companies and generic drugs sold in India, the repercussions of which didn't bear thinking about.

Throughout those years of the case, Novartis would repeatedly threaten to remove its pharmaceutical efforts in India completely, repeatedly reminding the world of its huge corporate power. It talked about how much research and development had gone into the drug, and how expensive it is to take a drug to market.

On 1 April 2013, the Indian Supreme Court rejected Novartis's legal challenge, representing a huge win for not only India's generic drug manufacturers, but for both Indian patients and those elsewhere in the developing world. The government noted that the pricing of the cancer treatment was arguably the most important factor in determining India's position in the case. With a monthly dose of the patented version of Glivec coming in at around $2600 per patient, it was over three times an average Indian's annual income.[60]

While it's true that developing a drug and taking it to market are indeed extremely expensive things to do, pharmaceutical companies still make huge profits compared to other industries. In 2019, the pharmaceutical industry was found to be the tenth most profitable industry in the US out of a hundred industries reviewed.[61]

Governments all around the world can and do have a say in how healthcare is delivered to their citizens. With better understanding of how the cancer therapeutics industry works as a whole, we citizens have far more power in petitioning and voting for better, fairer systems. If we are caught up in hyped-up headlines about unproven drugs, or believe that the pharmaceutical industry wields more power than it does, we run the risk of letting the status quo continue, with better cancer treatments possibly lost in the sands of bureaucracy and corporate greed.

Hype and How We Feel

Hype in the cancer world isn't just about corporate profit, the wrong drugs coming to market or misguided over-excitement. When people's length and quality of life are part of the conversation, narratives can also have painful emotional repercussions.

One fact about cancer is that people will still continue to get some form of it, and suffer incredibly at the hands of such a disease in the years to come. We can come up with new treatments and drugs and ways of managing the disease, but people still have to live with the fact that something inside them is killing them in an unpredictable and sometimes vicious manner. People will still suffer from cancer and its therapy's side effects, hopefully to a lesser extent as the years go on, but it seems highly unlikely we'll live in a world without anyone, ever, enduring cancer's agony. Cancer happens as a result of errors in the natural division of our cells; we simply cannot stop that from happening.

Hyped-up headlines can be cruel to patients. Suggesting a breakthrough, or even a cure, in the mainstream press, where people with the disease or their loved ones are likely to see it, can over-inflate hopes and ultimately lead to distrust in the system when their doctor tells them the drug is not relevant to them (or if it ultimately doesn't work after having spent huge sums of money). It can prompt patients and their families to pressure doctors to prescribe treatments that may not only not work but could cause extra suffering from side effects or disappointment.[62]

The rise in alternative approaches to healthcare outside of the medical system over the last few years seems linked to the lowering of trust in medical practitioners paired with the huge access to information – both good and bad – that the internet affords us.[63] Consider the crisis in trust around the measles vaccine worldwide. There are many cases of patients losing trust in their doctors and taking to unproven, possibly life-shortening methods promoted by those who stand to benefit from this distrust.

Fear Sells

Charities, companies and governments frequently use 'battle narratives'[64] when describing their work in cancer. Phrases such as 'we will beat cancer', 'the body is a battlefield' and 'fighting the alien inside' do well to prompt a sense of urgency in those from whom they are asking for donations, investment and support. These narratives are known to sometimes give patients a sense of empowerment and control in their 'own personal fight', providing a sense of purpose within those who are suffering.

But the violent narratives are also problematic in putting undue pressure and a sense of duty on to the suffering person. Sometimes, patients feel guilty when treatments don't work, that they didn't 'fight hard enough'. The common comment 'they lost their battle with cancer', to some, might suggest a lack of trying on the part of the patient. As Kate Granger, a terminally ill person, wrote in the *Guardian* in 2014: 'I would like to be remembered for the positive impact I have made on the world, for fun times and for my relationships with others, not as a loser. When I do die, I will have defied the prognosis for my type of cancer and achieved a great deal with my life. I do not want to feel a failure about something beyond my control. I refuse to believe my death will be because I didn't battle hard enough.'[65]

When the narratives are used to raise money or provide corporate profit, it can leave a particularly bad taste in the mouths of those who feel their own journeys are out of their control.

And it's not just the patients themselves who are affected by the way we talk about cancer. The feelgood stories presented in the media of 'miracle patients' who survived and 'battled the disease' against all odds make us think that if it were us, *we* would fight, *we* would be the one to survive, *we* would be that outlier.

The narratives and the hype around cancer may sometimes be empowering and positive, but they ultimately mask that thing many of us have buried deep, or maybe never even confronted: our fear of death.

That collective fear can fuel not only emotional toil but ultimately bad decision-making, which leads to cancer research, ironically, getting less attention and a less beneficial critique to help move it forward more positively. Our collective fear of death, and, in some predominantly Western cultures, its perception as a taboo topic we shouldn't discuss, contributes to our uncritical support of promising headlines and our broader lack of understanding of how to change the status quo. We don't know that we should demand better when we aren't confronting the reality of the disease, which requires facing the fear that cancer brings.

Fear sells, and sometimes it sells bogus goods.

As my understanding of cancer increases, and I understand more about how the research field is split up and the opportunities that lie ahead, with so many discoveries still to be made, I'm torn between feeling optimistic and pessimistic about the field. In some sense, understanding that cancer is just the splitting of cells makes it feel easy to treat somehow, but understanding the exponential variance that comes with trying to get to the bottom of each individual's cancer only makes me realise how far we have to go, and how unrealistic is the idea of a universal 'cure for cancer'.

And as much as that understanding can harbour a general sentiment of 'what's the point?', succumbing to that is just as problematic. With deeper understanding comes questions about the way we develop treatments, the way we regulate new drugs, the way our governments make decisions on healthcare; and those questions lead to better understanding of what needs to be done better across the field as a whole. More people understanding that, and taking action as a result, can only be a good thing.

Nothing changes if we don't, and we can change our way of thinking about cancer right now by challenging hype and in-built narratives presented in the media.

Hype in the cancer world ultimately plays on our fear of death, and that blinkers us from looking deeper, spotting the blockers to progress and knowing how to challenge the status quo.

If we can't 'beat' every form of cancer, at least let's try to beat the hype.

The Delicate Ethics of the Battery World

It turns out dead batteries have become a modern trope in horror movies.[1]

It used to be that the thrilling moment when your heart drops and you are certain doom is right ahead would be when the hero runs to the phone and picks up the receiver, ready to call the police or their friend – ANYBODY! – only to find the cord of the phone line cut and a silence ringing in their ear.

But nowadays, we don't really use landlines. Certainly not ones with phone cords. And so instilling an audience with the desired fear reaction means showing them a world in which they themselves can see no way out; and that is the world without their mobile phones. Indeed, in Jordan Peele's Oscar-winning *Get Out*, the main protagonist keeps finding his phone unplugged when he's trying to charge up the battery.

Beyond the possibly life-saving quality of having a fully charged mobile phone, we need better batteries for many parts of our modern lives. Yes, we want batteries that solve the problem of our phones dying right at the point we place the order on those early-access tickets. But we also want to find the battery that will make electric cars go further without having to stop mid-journey for a recharge. We all want to live in a world where renewables can effectively power our planet without the need for fossil fuels (yep, batteries play a part in the root of that problem too).

There's huge demand for better batteries, and thus a big opportunity for whoever can solve their problems.

And so often when a new design, a new material or a new method is announced by a battery startup or scientist, the battery is touted as perfect in that it checks the key boxes the battery industry is looking for: safe, cheap and able to store a lot of energy. When it is claimed a battery is able to satisfy these industry desires, you can put money on some headline hailing the arrival of the 'Holy Grail of Batteries'. Indeed publications and programmes such as *Forbes*, CNBC, *Utility Dive* and *Oil & Energy Investor* have used this exact phrase in their headlines about new battery innovations.[2]

The 'Holy Grail'. A mythical cup which endows the drinker with eternal youth turned popular metaphor to describe the almost-impossible-to-obtain dream prize, solving all problems at hand, in whatever field to which it's referring.

Metaphors are important tools in communicating complex topics. They are shortcuts that bring everyone up to speed on the background, allowing the person doing the communicating to 'get to the point' without getting lost in the weeds of explanation. They get rid of the need for a foundational education in every field; bring life and excitement into dull stories; and ultimately, with their emotive and memory-boosting qualities, make persuasion all the easier.

With language such as 'Holy Grail' being used to describe innovations in the complicated world of science and technology, you would be forgiven for thinking these innovations are perfect, the complete solution to all the problems. But just like the Holy Grail might bring an unintended curse of eternal youth, so too might these innovations have problems of their own. Maybe you have to watch each of your loved ones die over and over with each new era you live in, for example. Perfection is a difficult thing to argue against, particularly when technological progress is often presented as something that makes the world a better place. Maybe that technology will spare us time and money, or reduce our carbon footprint, or even save lives.

But herein lies the power of hype and idealism, and the metaphors we use to convey them: it blinkers us from seeing what's really going on and, with that, the answer to solving those grave problems we don't even know exist. Shortcutting thorough explanations by using popular metaphors might result in the message being lost. The shortcut might cut out the main points. And sometimes those shortcuts, those metaphors used to convey a complex sentiment with speed, they cut out arguably the most important thing to know: the nuance that comes with a complex message. In the case of our 'Holy Grail' metaphor, in using it to describe innovative science and technology, we miss out one crucial piece of information: why said 'Holy Grail' is not perfect.

'The Holy Grail of Batteries' makes sense as a sure-fire way of getting a reader to believe that an innovation is perfect, and therefore worth buying or investing in. In the case of batteries, though, there is simply no such thing.

BATTERIES IN DEMAND

It's no secret that electric cars are all the rage. They eliminate the need for petrol or diesel; they're getting cheaper with time, as the volume of cars manufactured and sold grows; and they give people the opportunity to align themselves ever closer to Elon Musk. (Musk will come up often in this book considering it's about new technologies and hype; he is the founder of many technology companies including Tesla [electric cars], SpaceX [rockets], Neuralink [brain–computer interfaces] and PayPal [online payments].)

Demand for electric cars is due to increase at an astounding rate. In 2019, only 2.1 per cent of cars sold were electric, and by 2040, it's estimated to be 32 per cent, according to reports from Bloomberg.[3] This huge predicted growth is subsequently heightening demand for their most expensive component under the hood: the humble battery.

The battery is what makes the cars so difficult to get to market. A battery that is light enough to not weigh down the car, as well as energy-dense enough to store the fuel to drive far without a charge, isn't cheap. And to keep the overall cost of the car low – in order to get people actually to buy it – manufacturers have to take a hit on their profit margin, reducing their incentive to even bother creating an electric car in the first place.

And this is just the car market. Think of the number of mobile phones and gadgets that have appeared in the world in the last twenty years, all in need of a lightweight, long-enough-lasting, cheap battery to keep them powered all day long. We've been in a bit of a battery boom, but that doesn't get away from the fact that our batteries disappoint us.

BATTERIES FOR RENEWABLES

The dream of having a planet powered almost entirely by renewables, such as solar and wind power, can only come to fruition if we develop batteries and systems that can store energy and release it at will. The problem that batteries can solve is 'intermittency'. We can't predict when wind will blow enough for the turbines to spin, and sun will only shine during the day for solar panels to gather light; but we need to be able to access energy at any time. Think about when you use the bulk of your energy. I'm willing to bet it's at night (when solar panels are useless), when you get home from work, turn on the lights and – in Britain at least – put the (electric) kettle on.

Electricity is not like oil, which can be stored easily in large pipeline systems. All electricity that is fed into the grid on one end must be used at almost the same time at the other end, otherwise the grid overloads and important infrastructure such as converters along the grid might get damaged. If we don't have cheap batteries that are energy-dense enough to store the energy created from renewables and intelligently distribute it at the times we really need it, renewables simply won't work as a viable source of the world's power. And to shake up an industry such as energy, you have to plug into quite the range of embedded, old infrastructure that not only runs throughout entire cities, but across the globe as a whole. Plugging in renewables that can't be controlled in terms of when they produce energy and when they don't is simply (from the perspective of the energy companies deciding whether to switch away from reliable fossil fuels) a waste of money. As we increase renewable energy efforts, we need to use the old infrastructure that allows us to pipe the energy produced from the plant to people who need it, at the right time; and so invent better batteries that allow us to store the energy produced and send it out only when needed.

And if too much energy is created and not stored, it's not a case of simply burning it off. California is known for producing too much renewable energy, and it's not only common for it to gift its excess energy generated from its solar panels to other states; it actually pays other parts of the US to take it.[4] Without good batteries to store that energy, California is at risk of overloading and damaging its power lines. In Germany, a different tack is taken when its renewables pump more into the grid than is demanded by its citizens: the price of energy goes below zero, and people are essentially paid to use it, as opposed to charged.[5] This loss of income through shrinking profit margins disincentivises energy companies from continuing to use renewables. We've still got so far to go if we want to replace fossil fuels globally, as at the moment renewables represent less than 30 per cent of global energy consumption.[6]

The only mature energy-storage technology that works economically at this scale is pumped-hydro – as in, hydropower stations attached to dams – and this is, of course, entirely location specific, as you need a good river.[7] Other technologies that aren't location-specific, whether they are larger lithium-ion batteries than you find in cars and phones, or alternative designs altogether, are simply still far too expensive to make sense on the grid.

THE TECHNICAL CHALLENGE

The quest for the Holy Grail of the battery innovation world is sometimes presented as simply the scientists not working fast enough to get new batteries into our devices and cars. The idea that there is 'one battery to rule them all' leads to confusion when the latest and greatest battery innovations aren't then put into the world's devices, cars and renewable energy stations. Indeed, battery startups have been popping up left, right and centre with weird, wonderful and genius approaches to 'disrupting' the battery problem, playing off the idealism that the problem can simply be solved with the spirit of entrepreneurship and invention.

For the lithium-ion batteries found in phones and cars, there are the scientists and entrepreneurs who are working on different materials to be used within the battery itself – to make it last longer and cost less. Some are substituting graphite for silicon, or substituting the liquid electrolyte (the medium that allows for the flow of electric charge inside the battery) for a solid one, for example.[8] Switching these materials, though, presents new problems such as batteries catching fire or materials cracking from within.

Then there are the scientists and entrepreneurs who are thinking up entirely new ways of storing energy at the home- or nationwide energy-grid level. Scaled-up lithium-ion batteries could be used, but they are very expensive to create, so while these batteries store energy chemically, many innovators are trying different tacks to get around their high cost. There are the flywheels – devices with high-speed rotors spinning inside – which store energy kinetically; or the sealed chambers containing hot rocks or some other material which keeps the heat, storing energy thermally; or the company doing pumped-hydro using underground shale reservoirs as opposed to dams; or the company choosing to use hydrogen instead.[9] These new ideas present many different kinds of problems such as how much energy they can store or how quickly the energy dissipates over time.

The overall problem of finding better batteries, though, is far more complex than simply 'inventing better technology'. To understand why we don't have better batteries, and why pushing hard for them can be problematic, it's not enough simply to understand the technical challenges that need to be overcome.

First, the wider industry, the business of the technology, needs to be taken into account. There are many considerations that impact on the way the industry behaves, such as: how difficult it is to get innovations out of the lab and into the real world; what incentives are at play for the companies, governments and customers involved; and, of course, what makes money and what doesn't.

Second, it's also the broader effect of new innovations on societies and their citizens. Who does it affect, and how? Who is losing and who is winning? What's happening to the planet in the process? Just because we want something doesn't necessarily mean we should prioritise that desire above the broader impact it may have – but understanding the impact that technology has can prove to be complex, contradictory and often opaque to the general public. All the more reason to try to explore what's going on in the background.

Understanding an industry is key to going beyond the hype. Getting to grips with the technology itself is important, yes, but it's vital to look at what's going on beyond the claims of inventors in order to put into context what is being said, and why.

THE BUSINESS OF BATTERIES

The first part of understanding batteries is knowing what's needed for them to go from the science lab to the market. This is not usually part of the chemistry curriculum, but is arguably the most important aspect to understand if the science is ever to have an impact on the world.

Different kinds of batteries serve different purposes. Some batteries are cheap but super bulky; some have huge storage capacity but are very inflammable; others don't last long but produce huge bursts of power; others are tiny in size but can't store much energy; the list goes on. However, a battery in a phone doesn't need to have a long life (as much as we may want it to), as cost is more important to those buying it when they know they can plug it in to charge. And a battery storing the energy for a large building, say, a hospital, can be as bulky as a shipping container as there's no need for it to be portable.

But it's not just about getting the right tech specs for the thing you want to power. There are whole ecosystems that batteries plug into. Take the lithium-ion batteries that you find in electric vehicles. The problem with these batteries, even though they are much better than most alternatives, is that they still don't hold enough charge for the electric car ecosystem to really work at scale. They need to be charged too often, meaning cities and neighbourhoods would have to start putting in enough charging stations to deal with the increased number of cars on the road. But building the charging stations is a big investment, particularly outside of cities where population density is lower, so deciding when to install and how many, before electric cars have really been adopted on a large scale, is a tough call. If the charging stations aren't there, though, people won't buy the cars.

The batteries, therefore, not only need to last for a long time, but also be as cheap as possible and not be too heavy as to weigh down the car.

The companies also need to decide whether they prioritise fast-charging if cities decide to have sporadic charging points or put their money towards helping towns and cities install more charging points.

It's not just about finding the best technology, but about finding the optimal mix of cost, size, energy storage capacity, rate performance (how fast you can charge it), and how easily it slots into the existing supply chain and infrastructure that the world is currently using to power homes, hospitals and everything else. It's not a case of 'one battery to rule them all', but rather about weighing up what needs to be optimised to ensure each battery is tailored for use for each application in the real world.

The concept of a 'Holy Grail of Batteries' therefore doesn't make sense. If the world of batteries is built on adapting to the use case, or optimising for different things at different times, the idea of there being a universal battery is, at best, misguided and, at worst, misleading. The idea simply doesn't match up with how the real world works.

Putting the Problem First

Not thinking beyond the technical question at hand, for example, working out what elements or set-up used in the battery is most efficient, can result in failure. Problems in industries don't just lie in the technology but also in the society in which they're used, the behaviour of the humans using the technology or even just the way people perceive the technology at hand. The Segway, for example: technologically brilliant in its clever balancing design, and greener than a car, it was touted as the next big thing in personal transportation. Unfortunately for the company's bottom line, the social consequences were huge. People using them were mocked, which put others off; the thing was too big and heavy to carry upstairs, so it wasn't exactly portable enough to ride in to the office and be parked next to your desk; and with the rider standing still on board, it wasn't even a healthy mode of transport that could compete with the bike. The technology was ace; the understanding of the problem of inner-city transport, and those who live with it, was not.

When new discoveries are made or new inventions created, there's a temptation to 'point' the tech at something and hope it solves a problem.

The issue with this approach is that in searching for a problem to fit a 'solution', confirmation bias is at play: 'I think this might be a problem, so let me go research why I am right,' as it were.

This is not how scientific research is done. Science is about finding out all that can be found about the world around us. When scientists investigate the world, they can't decide ahead of time what they're trying to do beyond simply finding out more about it. When it comes to then translating new science from the lab into the real world of society, culture, business and all the messy stuff in between, we must also take the approach of being open-minded about why the world is the way it is, and focus on what the problems of the world are first, before we try to shoehorn scientific developments into something ill-thought-through.

Technological idealism, and marketing hype, can miss the most important points when it comes to the real world – if you have inventors and researchers holed up in garages and laboratories not thinking about how the world really works, you end up with glorious, clever theoretical ideas and unfocused attempts at solving problems.

The key point often missed in the battery field is that there is a balance to be struck in finding the most optimal solution to a problem. And performing this delicate balancing act means working out what's most important to prioritise, depending on what the battery is being built for, and what can be put to one side. Only by understanding how the moving parts come together can we then start to invent and improve.

Sometimes the thing we optimise for is portability, sometimes it's energy density, and sometimes it's about finding the balance between factors. Most of the time, though, it comes down to cost.

Price, Beyond the Money

Despite the fact that there's a balancing act of size, energy-density, speed to charge and so on in choosing the right battery for a purpose, if a battery is to be adopted by a market, or replace whatever is being used right now, it also has to be cheap. Electric cars can't catch on with the masses if they're too expensive, and the battery is the most expensive part.

Renewable energy won't be adopted at scale if it doesn't make the whole energy system cheaper to run.

For lithium-ion batteries, the average price per kilowatt-hour (kWh) has been falling consistently over the last few years, alongside the increasing cheapness of most technologies. (kWh is the industry benchmark used for comparing batteries, as it defines how much energy the battery pack has available.) In 2010, it was $1200 per kWh; and in 2018 it was about $175, according to *Bloomberg New Energy Finance*.[10] Now, in order for electric cars to be affordable or at least similar in price to petrol-powered cars, meaning they will be more easily adopted by the mass market, it's expected that the price of the battery would have to be around $125 per kWh.[11] Indeed, if we continue at our current rate of progress, and with that, reduction in price, the electric car market is looking positive.

Great news for electric cars, but the same can't be said for storing energy for long periods of time to combat intermittency in the grid.

Batteries that solve the large-scale storage problems of the renewable energy industry are still far too expensive to make economic sense to invest in. There are many reasons why lithium-ion batteries found in phones and cars are too costly for large-scale grid-level storage. For starters, they use a modular set-up. One battery gives you one chunk of energy, two batteries gives you two chunks of energy, and so on. It very quickly gets pricey. The alternative approach to this modular set-up is the volume-based battery – as the amount of storage goes up, the cost per chunk of energy goes down. This works because it's more cost-effective to have one huge hydropower station with a dam, with one set of equipment and personnel and so on, than to have several smaller hydropower stations across the country, each with a set of equipment and personnel and so on. In short, they scale better.

Saying all that, it's been shown that using either lithium-ion batteries or pumped-hydro is prohibitively expensive to implement if 100 per cent renewables were to be used.[12] If California were to fulfil 80 per cent of its energy demand with renewables, it would need 9.6 million megawatt-hours of energy storage.[13] At time of writing, California only has 0.15 million megawatt-hours of energy storage. Looking at the US as a whole, if it were to meet 80 per cent of its electricity demand with wind and solar, it would require a battery storage system that would cost more than $2.5 trillion.[14]

Funding all this would mean raising energy prices – which would, of course, be passed on from energy suppliers to the consumers: us.

In order to make batteries a feasible technology powering the renewables revolution where we're not paying extortionate amounts to turn our kettles on, it's estimated the average cost of batteries would have to be only $10 per kilowatt-hour each – a far cry off the current average.[15]

COMPLEX COSTS

To understand why batteries cost so much, and why there's a limit as to how cheap they can be, it's key to understand that cost is not just made up of the price of the materials – there is a whole chain of costs involved in creating something from nothing.

Starting at the very beginning, there's the cost of the ideation (researcher's salaries, experimentation in the lab, the cost of machines and test materials); and then once something has been discovered, maybe there's the cost of regulation and government fees to have it approved for use outside the lab. Then, if a business decides to take the discovery and turn it into a product, there's the cost of licensing (buying the idea from the inventor); of doing more experiments to see how it can be scaled up to work in the real world, paying for machines and materials and user groups; of working out how to manufacture at scale – paying good engineers good salaries. Then there's the cost of mining large amounts of raw material (running the mine itself, paying staff and suppliers), of building and running the factory used to produce the final batteries (the rent, the building materials, the construction company, the energy bills, employee salaries). Finally, when it comes to selling the batteries, there's the cost of sales staff salaries and marketing materials; and then of shipping to the final customer – transport, logistics, fuel.

When you pay for a battery, this whole chain of events is taken into account when the price is set, and you're effectively paying the seller back for all the work that has gone into turning a load of raw materials into the thing you now own.

When a cheaper battery is touted by startups and the media, you have to work out in what way it's claimed to be cheaper, and whether the full pipeline really has been taken into account. If you see something being publicised as 'half the price' but it's not yet in the market, the reason why normally lies in working out at which point there exists a kink in the chain. This new battery might be using a cheaper material to mine, but it might also be a more expensive material to handle in the factory – maybe it's more toxic, for example. It might be a cheaper battery to produce in the lab, but it might also be more expensive to build it at scale – maybe the process of putting it together doesn't fit into existing factory lines, and whole new factories have to be built around it. Maybe the battery is cheaper in one country, but it's more expensive elsewhere: maybe the cost of shipping is high as the battery is huge, or it takes longer to go through customs because there is a new mineral to declare.

And are the mining workers being paid less than before? Is the factory in a country where carbon tax does not need to be paid? Is the battery worse for the environment?

The answer to all these questions might be no, but not considering the industry as a whole, and focusing only on the technical challenge, mean that when bad practice and corner-cutting does happen, it can easily be hidden in the hype.

Making the Switch

Understanding the business side of science and tech isn't just about getting to grips with costs – it's also about recognising the peculiar challenge of getting companies to change their behaviour.

Inertia is a hurdle often forgotten in the innovation pipeline, meaning industry stagnation and keeping to the status quo. Simply put, sometimes no matter how good a new solution is, how much time and effort it saves, how much better it is for the environment, people sometimes just don't like to change. It's too much effort.

It might be because change is costly. For example, many businesses still have their employees using old versions of Microsoft because it's expensive to have to redo the licence for everyone, it takes time away from employees' day-to-day work to spend time upgrading their computers, and there are no hard costs to leaving things the way they are. The cost of inertia is maybe a loss of productivity due to out-of-date software and slow programs, and maybe there is low employee morale, but the barrier to change is high.

From a business perspective, arguably the biggest mistake of the car industry over the last hundred years is not controlling the value chain around cars: not just manufacturing and selling the car itself, but selling the petrol and running the petrol stations and building the roads and inventing all the components under the hood. If the car industry had control of all of the 'bits' that go into the experience of getting from A to B in a car, and then decided to switch to making electric cars, surely we'd already have them on the roads in vast numbers (if we assume that the car industry wanted to act responsibly in the process). The car companies wouldn't have to negotiate pricing of components as they would be making them themselves. They wouldn't have to worry about where the charging stations would be, because they could use the network of petrol stations they already own.

Instead, car manufacturers and battery suppliers and town councils and consumers are all having to somehow nudge each other to change bit by bit – everyone at some point having to accept the brunt of a cost or invest time in rethinking routine – in order for the thing that's more or less universally accepted as better (in this case, the electric car versus the petrol or diesel car) to take hold in the market.

Making change happen can be hard not just from a cost or time perspective, but from a behavioural one too. Think about the effort involved when changing from, say, an Apple phone to an Android. Many people dislike Apple but feel they are 'in too deep' and the effort required to shift over to Android simply doesn't feel worth it.

Sometimes, people don't want to have to shift from what they've got used to, and companies are made up of people. Inertia in markets can sometimes be hugely to blame when the question arises, 'Why can't we just do X?'; and inertia can affect both companies and consumers: every one of us.

CORPORATE SHORT-TERM THINKING

Short-term thinking can also be to blame for inertia preventing better ways of doing things. Maybe an upfront cost takes precedence over long-term savings, for example. Perceived risk can also be a factor, especially if companies are worried about loss of earnings if they decide to make a shift they don't yet know for sure will pay off.

It's in situations such as these that governments, as opposed to corporates, make for the best catalysts. They can put in regulations that force change; they can offer subsidies for businesses to incentivise them to take risks; they can give out grants that don't require any payback to fund work that is developmental and not yet making any revenue.

In the West, batteries are predominantly treated like a business and engineering problem – meaning government doesn't get all that involved – and so market demand drives change (or lack of change) in the industry.

Beyond supporting early-stage science that shows potential and needs help getting to the point of commercial viability, another reason a government might get involved in tackling inertia in technology adoption is if it predicts that success in the industry will build some kind of home advantage. Maybe it would improve health outcomes for citizens, or maybe it would tackle issues with the environment in that country – or maybe it would build geopolitical and economic power beyond any other nation on Earth.

There's one country that understands this final argument all too well, and is charging ahead in the race to lead the world in the development of batteries.

CHINA'S ELECTRIC CAR REVOLUTION

In 1990s Shenzhen, a little-known batteries company called BYD was formed.[16] With their fast processes and cheap production methods, they were soon making batteries for Nokia mobile phones, Dell laptops, drills, digital cameras and anything else in which you could find the increasingly popular lithium-ion battery lurking.

As the world demanded more batteries, BYD's revenue skyrocketed, and with their successful IPO on the Hong Kong stock exchange in 2002, the management decided to buy a failing Chinese state-owned automaker, and pivot into the world of cars. In 2008, BYD launched the world's first plug-in hybrid car; a few months later, Warren Buffett acquired 10 per cent of the company. BYD now sells around 30,000 plug-in hybrid or pure electric vehicles in China every month. China accounts for more than half of worldwide sales of electric vehicles (in 2017, there were 580,000 of them sold in China, the next country on the list was the US, with just under 200,000); the city of Shanghai had more electric vehicle sales in 2018 than France, Germany and the UK combined.[17] The company has about 250,000 employees, is the world's biggest manufacturer of electric cars, and has expanded into various other areas such as buses, solar panels and even monorails.

But it wasn't just savvy business building that made BYD such a success. The Chinese government has pumped billions of dollars (between 2009 and 2017, around $58 billion, to be more precise) into electric vehicles – in subsidies for those buying them, in infrastructure investments such as charging points, and in research and development grants.[18] Restrictions were also put on those buying petrol-powered cars: in Shanghai, in order to register your new petrol-powered car and get a licence plate, you need to enter a licence plate auction, where prices are at around $14,000 (never mind the price of the car itself).[19] A licence plate for an electric vehicle is free, and BYD's cars cost only around $9,000 after subsidies.

The Chinese government's actions make sense from a domestic perspective: those who live in China suffer from terrible air quality due to pollution, and much of the crude oil that China imports comes from the Persian Gulf and thus has to travel through waters with the US Navy ever present.[20] The push for electric cars helps clean up Chinese air and reduces reliance on international oil, while it continues to build the Chinese economy.

And it's not just BYD selling electric vehicles to Chinese consumers. There are around 400 electric vehicle makers in the country, and foreign car manufacturers are wanting a share too. Volkswagen is planning on introducing thirty new electric models in China, manufacturing them in its new Shanghai factory.[21] Tesla is also now building its own plant in the same city, in a bid to capture some of the Chinese market and perhaps finally begin to become profitable in the process. At time of writing, Tesla has recorded only two profitable quarters in its history, while BYD has always been in profit in its transition to electric cars and growth.[22]

BEYOND CHINA'S SHORES

Beyond China's borders, electric car sales from Chinese brands aren't so impressive. With decades-long narratives around Chinese-branded consumer goods being of lower quality, and privacy concerns surrounding brands such as Chinese phone manufacturer Huawei, the Western customer tends to avoid Chinese brands.[23]

The batteries that go in the cars that Western customers are more willing to buy are a different story.

Enter CATL – the largest lithium-ion battery manufacturer on the planet, based in, you guessed it, China.[24]

CATL already supplies the batteries for more than 40 per cent of all the electric vehicles in China, but its sights are set outside the country as it looks to grow. In Germany, CATL is building a factory to be in close proximity to the likes of Daimler, BMW and Volkswagen (who, by the way, are already customers). It has built offices in France, Canada and Japan; it's bought 22 per cent of Finnish company Valmet Automotive (which supplies batteries to Mercedes-Benz, Lamborghini and Porsche); and it has signed a multimillion-dollar battery supply deal with Sweden's Volvo.[25]

CATL's growth plans – if they materialise – mean that China will be supplying 70 per cent of the world's lithium-ion batteries by 2021.[26]

The electric vehicle revolution isn't just about cars, though: electrifying buses, trains, trams, lorries and planes is arguably more beneficial in terms of reducing carbon emissions – and these huge transportation industries are in need of efficient, light, cheap batteries too. Both CATL and BYD – as well as many other Chinese electric vehicle companies and suppliers – can be found in transport systems all over the world. China, in fact, makes 99 per cent of the world's electric buses – most of which are found in China.[27] But clean energy regulations gradually coming in around the globe mean demand for electric buses is going up. The government bodies tasked with buying buses tend not to care where they come from, as long as they're reliable and cheap, and even when governments stipulate that they must buy vehicles manufactured or assembled in their own countries, BYD got around this by simply building a factory in California.[28] Individuals who use those buses every day tend not to care, or even know, who made them or where they came from.

Even the electric scooters taking over cities across the world – with rechargeable batteries inside – are mostly made in China.[29]

China's Bigger Vision

China's activity in the battery world makes perfect sense when it's put alongside their strategic international supply play. China isn't just building incredible technical capability domestically, and operating battery factories elsewhere, to sell to the rest of the world. It is literally building entire new infrastructure the world over and, in the process of doing so, gathering up control of the supply chain and energy systems across the globe.[30]

To meet international demand for Chinese goods, batteries included, the Chinese government created what's called the Belt and Road Initiative: the strategy involves infrastructure development such as roads, hydropower dams, rail transportation and ports for sea trade routes all across Asia, Europe, the Middle East, Africa and South America. China is investing around $4 trillion over the course of the project, due to be completed in 2049, with over 600 projects already on the go.[31] With that kind of investment, both business and power are there for the taking.

Almost half of Cambodia's electricity comes from seven dams that China built and financed.[32]

Sri Lanka failed to pay back Chinese loans, and so they handed over control of a crucial port to settle the dues, prompting many to accuse Beijing of catching partners in debt traps in order to seize assets.[33]

While Greece was answering to the rest of Europe for the repercussions of its 2010 financial crisis, China invested in its Piraeus port and turned it into the busiest shipping hub in the Mediterranean – and now a gateway to the rest of mainland Europe.[34]

At the end of 2018, there were 41 pipelines delivering oil and gas; 203 bridges, roads and railways connecting China to the rest of the world; and 199 power plants – across 112 countries around the world – all paid for, built and owned by China.[35]

In short, China is looking to gain dominance in almost all underlying power projects worldwide; and the battery industry is just one of them. When we look at the battery industry, we must consider it in the context of this wider power play, and not simply as a technology one country happens to dominate.

So, understanding the business behind battery technology leads us to complex costs, no 'one size fits all' approach, and the growth of power in China. It's not enough to invent a new kind of battery that holds a little more energy, or costs marginally less to manufacture, or is a fraction lighter. There is so much more that goes into getting the technology into the phones, cars and power stations of today. It's easy to believe that the battery market is revolutionised when one company or scientist claims to have found the 'Holy Grail' in terms of the tech specs, but better science is simply not enough to go on. Unless they also explain their full cost analysis, or their global distribution plan, or their method of getting all the incumbent device, car and power-plant makers to adopt their new technology over their current (probably Chinese) suppliers, the Holy Grail narrative is most likely a hyped-up way of explaining the value of their work, without full consideration of the market in which they must exist.

WHERE DO BATTERIES COME FROM?

The second thing to get to grips with in understanding why we don't have better batteries (and why pushing hard for them can actually be problematic) is what the broader societal effects are when we demand more, better, technology.

Lithium, cobalt, nickel and manganese are the main elements required to manufacture lithium-ion batteries. All of these elements are mined in problematic and unsustainable ways.

Cobalt is the most concerning. Sixty per cent of the world's cobalt is mined in the Democratic Republic of the Congo. There are reports of child labour; dangerous work conditions (including 100,000 cobalt miners having to use hand tools to dig hundreds of feet underground); frequent death and injury; and exposure of local communities to toxic metals, which seem to be linked to breathing problems and birth defects.[36]

In 2017, the UN found mass graves, which it blamed the Congolese army for digging, and linked them to both ethnic rivalry and competing claims on the country's rich mineral resources.[37]

Around 90 per cent of China's cobalt comes from the Democratic Republic of the Congo. The *Washington Post* traced the cobalt along a lengthy chain dominated by Chinese interests.[38] First, the cobalt moves from the small-scale Congolese mines to one single Chinese company called Congo Dongfang International Mining, owned by one of the biggest cobalt suppliers, Zhejiang Huayou Cobalt. It then sells it to various battery companies around the world, eventually ending up at the companies – such as Apple – creating the technology we demand with batteries inside.

It's hard to come up with an alternative to cobalt – it boosts battery performance unlike anything else – meaning the companies making computers and cars are fuelling a supply chain rife with questionable human rights practices and governmental corruption.

The electric car market is what's most troubling for those concerned about the cobalt market – a phone battery contains only about 10 g of refined cobalt, whereas a single Tesla battery contains 4.8 kg.[39] There are efforts being made to reduce the cobalt content of automotive batteries, with suppliers such as Panasonic shifting to more sustainable designs.[40] But the need for the element is still there, and with electric vehicle consumer demand rising, they are projected to make up two-thirds of the demand for the precious mineral.[41] Questions therefore are rightly being asked about the mining practices of the most expensive central element in the batteries of our phones, laptops and – soon – vehicles.

The questions are not being asked loudly enough. There are four minerals considered 'conflict minerals' – tungsten, tin, tantalum and gold (commonly referred to as 3TG) – and due to the mining of these materials being proven to contribute to war, both the EU and the USA have regulation surrounding their importation.[42] Cobalt, however, isn't considered a conflict mineral – it's considered an unethical but not a conflict initiating mineral – so the onus is on the companies to make responsible decisions about where they get their raw materials, outside of formal regulation, and at the expense of their own profit, of course.

Thinking only about the Western companies that make electric cars or smartphones is failing to see the bigger picture. The battery is the most important component of these devices and these companies have little control, or at least exert little control, over where they come from. By quietly avoiding cleaning up their supply chain, they reap the benefits that cheap materials have on their reported profits, and still are lauded as some of the most innovative and inspiring companies to grace the Earth.

Those supplying batteries, in other words, Chinese companies and the authoritarian government behind them, have huge power. If we get caught up in the 'cool factor' of Apple or Tesla, and forget to look beyond the brand into the underlying systems powering the companies, we run the risk of failing to scrutinise what is going on behind the scenes.

China doesn't just control the battery market through its huge companies and running the cobalt supply chain. China also owns 50 per cent of the world's lithium mines.[43] In 2010, at the height of a dispute between China and Japan surrounding a Chinese boat captain who had collided with a Japanese coast guard boat in long-disputed territorial waters, China halted exports of so-called rare earth elements.[44] Japan eventually buckled and released the captain back to China. China's huge advantage – which is still growing – in the minerals space makes for great leverage in any future trade wars.[45]

SMART COMMUNISM?

It's not just that Chinese companies have a 'sixth sense' in knowing which global infrastructure areas to invest in before the time comes for them to really make their mark on the world. The Chinese success in the battery arena comes down to the fact that the Chinese government makes long-term decisions, sometimes at the expense of immediate cashflow, in order to keep the country growing.

In 2007, the biggest financial crisis since the Great Depression in the 1920s took the world by storm. Unemployment shot up, banks had to be bailed out across the globe, and the European Union's internal relationships became imbued with resentment when countries had to prop up those that were failing. It would mark the beginning of the era of austerity, which would last for more than a decade.

The story isn't quite the same in China.

As its economy started to plunge from the slowing demand for exports, the government decided to plough $568 billion into the economy, to prevent it from failing further.[46] It was the biggest economic stimulus ever undertaken in China. The government invested in railways, roads and power lines. It ordered the state-backed banks to give out loans and state-backed companies to keep investing and building.[47] It put limits on how much money its citizens could move out of the country, keeping much of it inside China.

A year after the announcement, Chinese consumers were spending enough to keep fuelling the country's economy, whereas elsewhere in the US, Europe and beyond, consumers were holding back for fear of further financial downturn.[48]

In taking this authoritarian approach, Chinese companies were able not just to survive the global financial crisis, but to thrive in the immediate aftermath of the crash.

Compare this to the Western capitalistic approach, where progress and innovation are largely market-driven, and governments count on businesses and consumer demand to forge successful investments in new ideas. Markets aren't necessarily great fortune-tellers, though, and businesses tend to make short-term decisions, especially when faced with impatient shareholders, poor cashflow and demanding consumers ready to shop elsewhere for whatever service is provided. Huge upfront investments in factories, in financing and in experimentation don't happen when the market doesn't allow for it, or when there's not enough money in the corporate bank account.

Not in China, where the government can dictate the long-term strategy of the economy and those within it. This is not to say China's economy is perfect. There is debate about the success of its approach to debt, about the productivity of government-owned companies versus the privately owned ones, and the cost to the environment of its rapid industrial expansion (China also builds hundreds of coal-fired plants alongside its investment in batteries).

Weighing Up China

In the West, communism tends to be seen as a dirty word. Understandably so, when it's viewed through the lens of brutish dictators and its impact on human rights. The idea that China and its authoritarian approach is on the road to controlling energy systems around the globe is a frightening one. This is despite the reality that China is really the only country capable of meeting the rest of the world's sustainability demands, such as the number of electric cars to be on the road, or the percentage of energy sources to be renewable.

It prompts us to ask ourselves the uncomfortable question: is having an authoritarian government with that kind of underlying power necessarily a bad thing?

The question as to whether China's dominant role in being the supplier of batteries to the world is good or bad comes down to what we're optimising for. If, as a global society, we made the decision that curbing use of fossil fuels was the most important job to do right now, we might also decide that it would be irrelevant which country, and by which means, was at the heart of the push towards a battery-powered planet. The short-term negative effects would be 'worth it' for a faster route to a greener Earth.

If, however, as a global society, we decided that human rights in the year 2020 must be upheld around the entire planet, and all efforts must be pointed to curbing any failings on that part, then it could be argued that the environment might have to take a hit.

In some sense the two are opposing players in a zero-sum game while we await another solution to the problem – such as an alternative to cobalt, for example. If you're to pick sides, you have to weigh up current civilisation's human rights versus future generations' living environment: an impossible choice for the majority.

MORAL PHILOSOPHY AND BATTERIES

So, we're left in a sticky situation: our demand for batteries is ever increasing, both in terms of 'saving the planet' as well as building cheaper, more efficient systems, while that demand simultaneously puts huge pressure on human rights.

In *The Good Place*, a television series, a dead woman ends up in heaven by accident. Her points record (the means by which the goodness of humans entering the afterlife is measured) is that of a careless human being, not a saint, so she enlists a professor of moral philosophy to help her 'become good' so that she can stay. Over the course of the series they get to see how the points system works, and they realise that for every point-scoring good action there are also negative points. For example: rehoming the last three dogs from a rescue centre, as opposed to only taking one, would garner lots of points, but the fact that the little boy who came into the pound afterwards is denied a dog would mean points are deducted.

So in order to come to some kind of conclusion about whether our demand for batteries is a good thing, we need to find a way of understanding the web of impact, and how that impact works, around the world of batteries. And without being overloaded with information, and then decision paralysis, in the process.

There's a balancing act when it comes to advancing the state of the world – and in the case of the battery industry, it's a delicate mix of climate change, human rights, communism versus democracy and geopolitical control.

A RACE AGAINST TIME

Finding a battery that works for the many doesn't just mean trying to invent one that can satisfy all the technical requirements of energy density, size and weight, and safety. If the 'Holy Grail of Batteries' could be created, it would also have to be perfect in terms of its effect on geopolitics, human rights and market adoption on a mass scale.

One universal solution simply doesn't exist so, instead, we need to work out how to distinguish between what constitutes a good decision and a problematic one. Or rather, knowing that all of the decisions will have some kind of problematic element to them, how do we rank them in order of how 'good' they are? Where does each option sit on the morality scale?

There's another lens through which we can weigh up our options: time. We're in a climate crisis, and one of the main drivers behind the move towards electric vehicles and renewables, both of which require batteries, is that time is running out to get away from our reliance on fossil fuels. If we don't take more action now to battle the climate crisis, we'll soon cross the threshold of global warming and be unable to slow the predicted negative effects.[49] In short, we have to act fast to find a way out of our reliance on natural gas, coal and crude oil.

Now, if we wanted to start lessening our dependence on fossil fuels immediately, such as by using only renewable sources of energy, i.e. wind and solar, we'd have to accept that we don't yet have a cheap enough battery available to get past the intermittency problem, so we might have to take the hit in spending much more than companies and governments can realistically afford while we wait. If we choose something that can be implemented less expensively, we might need to pick a hybrid solution, thus not going for the full-scale renewable approach and maybe mixing renewables with nuclear energy or some fossil fuels. We'd also have to consider return on investment over time – how long would it take, say, an energy company to make back their investment in these expensive batteries for renewable storage, and is this a short enough timeframe that the company can continue to operate, without going bust, in that gap between investment and monetary return?

There's also the timeframe to consider in terms of who – and when – it impacts. Human rights exist in a very current timeframe; the protection of our planet considers a far-off timeframe, affecting many generations yet to come.

There's no perfect solution, so the question remains: how do we pick a track, know the downsides before we begin so we don't veer off course midway and lose the time already invested, and justify which things we're optimising for in our decision?

LET'S PLAY A GAME

Imagine you have a bag of ten credits. You get to assign your credits in whichever order and with whichever spread you choose across three areas: climate change, human rights and geopolitical stability. If you put all ten on one area, all the innovation and future decisions around the battery industry go towards optimising for that best outcome. If you split the credits three-three-four, then you make decisions that essentially compromise each of them, and so on.

Thinking about what's 'best' for the industry is like a 'build your own adventure' game where, at every turn, each decision is made by you based on your interests, morals and view of the world. There are many different outcomes that can happen depending on where you place your 'credits' and therefore which decisions you make, and any of them could be argued as morally right depending on the culture you come from and the timespan you work within.

Which leaves us with the realisation that this is why this industry is in the state that it's in: there isn't a 'right' answer. But therein lies the fascinating thing about exploring an industry when you consider all the angles. There's a feeling of inertia; that everyone is pulling and pushing in different directions based on their own views about what is best, leaving the industry as a whole to kind of 'average out' into nothingness – or rather, a slow but steady form of 'progress', with downsides we tend to avoid discussing.

In order to get out of this state of incremental progress, maybe a big shift is required. Maybe it comes in the form of a piece of legislation, or a PR campaign that changes public sentiment, or an invention that changes the nature of the available technology. Either way, it requires a level of short-term irrationality with regard to capital, or a level of bravery in deciding what the compromise is.

Let's pick option one: short-term capital irrationality. Consider the cobalt problem. Alternatives are expensive, not as energy dense and increase the weight of the final battery, but the mining of cobalt as demand increases looks to raise not only humanitarian but geopolitical issues. Let's say Apple made the decision that it wanted to put seven credits on human rights, and stipulated that it would now only use certified cobalt, much like free-trade bananas, in the manufacturing of its devices. To begin with, this would cause huge problems for Apple. The supply chain isn't set up right now to meet those kind of demands, meaning the prices of its devices would most likely have to go up, losing the company customers and therefore profit. Or, it would have to take the hit in order to keep the customers, and still lose profit. In the short term, this could be deemed irrational, particularly by shareholders if they haven't had assurance that the supply chain will adapt in time to prevent Apple going bust. But, as with other such radical decisions made by powerful industry giants, the market tends to adapt around them eventually. The question is simply at what speed, and at what cost to Apple's bottom line, should change occur? Of course, with more advocacy against questionable cobalt, more customers will shun Apple products (you would hope) in a similar vein to the way in which some people boycotted Primark due to sweatshop manufacturing. And if you get the right critical mass of consumer disgust, companies do change. Again, though, it's all about timing, and who's willing to be brave enough to be irrational in the short-term.

Moving onto option two: bravery in deciding to compromise something. Let's decide that we're going to put our ten credits on climate change, and make the decision that we're happy for authoritarian regimes to forge ahead, for the good of the rest of the planet in the years to come. Geopolitics and human rights would then be 'taking the hit', but just in this decade while the rest of the countries catch up with China in terms of battery-manufacturing capability or while scientists work on an alternative. The compromise here is pretty shocking, right? But you have to consider that this is the sort of decision that is essentially being made through lack of attention: by not considering human rights, we're essentially putting our ten credits somewhere else. By stepping up and making a decision, forging forward with something in the hope of getting there quickly while trusting that someone else will solve the other problems concurrently with their own ten credits, maybe that bravery, with all the social rejection and moral compass considerations that go with it, is what is needed.

This credit system, merely a thought experiment, doesn't mean that working out solutions is always a zero-sum game. The point here is that without considering the balancing act at all, certain areas lose out. When trying to work out solutions, it's key that full consideration of all the elements of the problem are brought to bear, and there's an active choice to be made around which areas are being focused on, as opposed to simply announcing that everything is going to be solved with one magic fix and leaving people, the market, the environment, and ultimately the shareholders, disappointed. Some innovations can solve various problems to various extents, but it's highly unlikely they can fix everything in one fell swoop, especially when certain elements in the balance directly oppose one another.

All of this assumes that if we were only to find some kind of ethically, economically and geopolitically sound suite of batteries that can 'plug into' the electric car, renewable energy and technology device markets, all would be well. But there's an argument to be made that even the very idea of creating a better battery is the wrong route to be going down.

Batteries themselves can be bad for the environment.[50] Extracting lithium harms soil and contaminates the air and water supply around the mines. Fewer than 10 per cent of batteries are recycled and the toxic chemicals inside tend to leak out when they are sent to landfills, or fires are started when they've been improperly stored or accidentally put through the crusher. Batteries degrade over time, so recycling isn't just tough in terms of consumer behaviour change, but also in terms of the viability of second or third use.

By increasing our use of batteries now, in order to curb our usage of fossil fuels in the car industry, make renewables more viable, and continue using our mobile phones and laptops, we're essentially creating another environmental problem in its wake. It's possible that in the years to come, we'll look at batteries as we look at plastic bottles today: the solution to one problem, but the advent of an even greater one.

A Balancing Act

Understanding the world's biggest problems is always about coming to terms with the balancing act behind them. Balancing the environment, human rights and war; balancing time to market with the cost of moving too quickly or too slowly; balancing the right technical make-up for the right use, and so on. It's a balance of science, business and ethics.

There's a good and bad side to most things. Missing that crucial truth is dangerous, as it not only blinkers us from the reality of getting new technologies to market, but also shields us from the problematic – sometimes immoral – 'other side' of the story.

It's not about branding a technology either immoral or perfect, but about being properly informed. And with that more informed approach, we are all able to demand better from the companies we buy from, nudge our behaviour towards being better aligned with our own morals, or simply have a better grasp of how the world really works.

When you understand the balancing act, it's easier to see where the change can then happen. If there are flaws in the system when it comes to the sentiment of the masses, we can work on marketing. If there are flaws in the system in terms of one country monopolising, we can appeal to government and regulators. If there are flaws in the system with regard to human rights, we can boycott.

Only by understanding how all those elements fit together – how they precariously balance one on top of the other – can we then point our energy in the direction on the most pressing challenges. If we are caught up with the smaller technical problems that – when solved – advance the science but don't fit in with the bigger system, we've no chance of changing the world.

PART TWO

Next
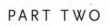

Belief and Disbelief in the Power of Fusion Energy

'Unlimited fuel for millions of years', reported the *Daily Mail* in January 1958; 'Something to wave the flag about', said London's *Daily Herald*; 'They have built a man-made sun on Earth', the *Daily Sketch* tabloid proclaimed.[1]

A flurry of headlines brought excitement and anticipation of a glorious energy future in Britain, after a press conference in England led the world to believe that scientists had managed to achieve fusion energy. Governments all over the world had been secretly working on fusion energy throughout the Cold War, and the UK's 1958 announcement came at a time when nations were keen to prove dominance in science and technology efforts. Indeed, the fusion machine unveiled at the Harwell research centre in Oxford that day quickly became known as 'Britain's Sputnik' and the rest of the world turned their attention to the small island making big claims.[2]

Unfortunately for the Harwell scientists, the UK and – ultimately – the fusion research community as a whole, they got it wrong. Their machine hadn't produced fusion after all.

The machine was called Zeta, for Zero Energy Thermonuclear Assembly, and it was built at the Atomic Energy Research Establishment in Harwell, just outside Oxford. The goal was to recreate the way energy is created in the sun – a process known as fusion – and the research across the globe

was competitive and predominantly kept classified due to the nationalistic sentiment of the Cold War era.

The 'Zeta fiasco', as it would come to be known, started with the leaks
to the press in 1957 that fusion had been achieved. The secretive nature
of fusion research meant understanding what was going on behind the
scenes was a difficult task for both experts and the general public alike.
Then, after the leaks had been speculated upon for a few months, the
team at Harwell opened their doors to explain their work. The conference began with a cautious statement from the head of the Zeta
programme, John Cockcroft: 'Neutrons have been observed in about the
numbers to be expected if thermonuclear reactions were proceeding.'
But after much questioning from the media as to what that meant,
Cockcroft said he was '90 percent certain' that Zeta's neutrons came from
fusion reactions.

Not long after the bombastic headlines spread overseas, the questions
from the US and USSR began, but with the cloak of Cold War sentiment,
they were dismissed as jingoism. Less than four months later, though,
more tests were run, and the Harwell team had to announce that the
neutrons they had observed weren't the result of fusion after all.

The Zeta fiasco was not only an embarrassment, but it proved to be a
blow to the credibility of fusion research as a whole. The public lost faith
in the prospect of recreating the sun. The idea that fusion was soon to be
demonstrated became a joke: for many it became the science that will
always be fifty years away. For others, it wasn't even believed to be real
science any more.

The difference between the Cold War race to prove humans could
land on the moon and that to prove fusion energy was possible on Earth
is that one big public display spectacularly happened, and the other spectacularly failed. One was hailed as the greatest feat of engineering, the
other was branded doomed to fail forever. The historic narratives of both
have held up. For those working in the space industry, the 'we can do it'
narrative has worked to their advantage, with it fuelling the inspiring
communications around adventuring beyond our planet.

But for those trying to create the one thing that could revolutionise
the entire planet's energy resources, which could, in turn, get rid of our

reliance on fossil fuels, their work and vision for the future have been not only tarnished, but actively restricted by that potent, false, narrative of the past.

Finally – fusion is coming. But only if we believe it to be true.

What Is Fusion?

Fusion is exactly that: the fusion of two atomic nuclei together to make a bigger one. The reason scientists are desperately working on it is that this process releases staggering amounts of energy.

The first thing to know about fusion energy is that it definitely works: it happens in the sun. Hydrogen nuclei smash together at tremendous speed, colliding with such force that the two nuclei fuse together and create one nucleus of helium. The energy released in that fusing process fuels our planet 93 million miles away; in short, there's a lot of it.

The sun's huge amount of gravity and sheer size mean that the hydrogen nuclei travel at extremely fast speeds, enough to overcome their positive charge, which would normally mean that they repel each another when they get too close. On Earth, it's harder to engineer. We don't have the benefit of a huge gravitational pull as the Earth's mass is so much smaller than that of the sun, so instead we have to work out a way of speeding up the nuclei so fast that they fuse and release that desired energy.

We typically use isotopes of hydrogen, deuterium and tritium as the fuel, and to get them travelling at those superfast speeds, we heat up the hydrogen isotope gas. The temperature needed for fusion to happen, it turns out, is over 100 million °C.[3] That superhot gas cloud at the right temperature becomes what is called 'plasma', and due to the heat being so high, it cannot come into contact with any solid material. The fusion reactor needs to be engineered such that the plasma can be confined safely for long enough at a high enough temperature and density for fusion to happen. Once it 'ignites', it can become self-sustaining, and the huge amount of energy put in to heat the gas up in the first place can be switched off.

Heating a gas to that temperature on Earth, in a contained manner so as to gather the energy created from the resulting fusion, is an extremely difficult task (nuclear weapons produce the unconfined and uncontrolled release of fusion energy). There are two main focus areas in the fusion research community that differ in how they confine the plasma. The more popular method is magnetic confinement, where very powerful magnets are used to keep the plasma in a vacuum away from the reactor walls and held compact enough so that the collisions can happen. The other method is called inertial confinement, where a high-energy laser or beam of ions is shot at a tiny pellet of frozen hydrogen about the size of a pinhead, heating and then compressing the pellet essentially to create a tiny explosion.

Both methods have demonstrated fusion, but the amount of energy required to make the fusion happen has been more than the energy that came out the other end, defeating the purpose. The point where the output tips over into being equal to, and then more than, the input is called 'breakeven' and it's that which the fusion research community is on a mission to prove can be done on Earth.

It can't be underestimated just how much of an impact human-made, controlled fusion would have on the planet. It's a source of energy that produces no carbon, nitrogen or sulphur emissions; fusion reactors aren't vulnerable to the dangerous kind of runaway reactions that happen in nuclear fission accidents; we have plenty of those hydrogen isotopes to hand, as we can generate them from ocean water; and for about 1 gram of that fusion fuel, it's been estimated we can create energy equivalent to that of about 7 tons of oil (enough energy for the average person over the course of a year).[4] It's too simple to say it would completely overhaul all our energy systems but, at the same time, being able to create more energy than we put in, in a much cleaner and greener manner, could be the answer to our fossil fuel concerns. Our world of electronic devices, connected homes, complex financial systems, advanced medical machinery, and all the other modern effects those in developed countries have come to expect as the norm, is desperately in need of a better way to power it all. As developing countries are coming online at a rapid rate and are adopting even higher volumes of technology, the need for a sustainable source of energy becomes even more apparent if we also don't want to rid our planet of all its resources and burn up our atmosphere in the process. As we found in the batteries chapter, we can't do it with renewables alone, so fusion presents a tantalising alternative if we want our planet to keep progressing at speed.

Fusion isn't perfect, as we'll see. Saying that, and bearing in mind that this book is about curbing hype, I don't think it's an overstatement to say that if we can find a way of getting fusion to work, at scale and within our current energy systems, it would truly be one of the most incredible feats ever accomplished by humankind.

What Fusion Is Not

Fusion isn't the same as fission – which is what current nuclear power stations utilise in their power creation. Instead of the fusing of two small nuclei into a slightly larger one in the fusion process, fission energy is all about taking one large nucleus (such as can be found in uranium or plutonium), breaking it into two and collecting the resulting energy emitted. The nuclear power stations of today are very controversial due to their reliance on and production of radioactive materials as well as the effort and cost required to create their starting materials of uranium and plutonium. Fusion power stations of the future largely rely on the relatively abundant input, water, and although some radioactive waste is produced, it's far less problematic to handle than that produced in fission power stations.

Fusion isn't the same as so-called 'cold fusion': hypothesised fusion which would happen at or near room temperature. In 1989, two researchers Martin Fleischmann and Stanley Pons reported that they had observed nuclear fusion in their glass jar apparatus at room temperature.[5] Their claims gained huge media attention, but after months of the scientific community trying to validate their claims without success, the consensus reached was that there had been experimental error on the part of the scientists.[6] Mainstream funding and support was withdrawn from cold fusion and the term became synonymous with junk science. There are, however, still some researchers and companies working on cold fusion, despite its muddy history (the field has been re-branded LENR, low-energy nuclear reactions).

When the public did listen to those reasonably calling out cold fusion as ridiculous, it created an incorrect narrative that fusion as a whole sits firmly in the realm of the 'crackpot inventor' types. To this day, many people confuse serious fusion research with cold fusion pseudoscience due to this lasting, but fascinating, narrative.

FUSION'S BEEN BURNED

It's no wonder, then, that little is said in the mainstream about modern-day fusion research. The overpromise of Zeta caused widespread scepticism due to its high-profile false start. The confusion between fission and cold fusion isn't abnormal – many people I mentioned this chapter to asked me why I was covering nuclear power when it's so controversial, or why I was bringing up pseudoscience.

The most damaging narrative, though, if you manage to reconcile Zeta, fission and cold fusion distractions, is that fusion is still so far away. The 'fusion is always fifty years away' cynical quip, which never seems to go away when fusion is brought up in formal discussions, reduces faith and trust in society's ability to make a fusion-powered Earth a reality. The history of fusion is often painted as a series of claims that fusion was around the corner, followed by failure, but the reality is far richer and with more nuance than this reduction gives credit for.

Repeating, and believing, reductive historical narratives damages trust moving forward, and may possibly slow or halt the development of fusion forever. The potential of this technology is far too great to be thwarted by flippant disregard based on what's happened before – we must therefore put the record straight.

Fusion is no joke.

THE HISTORY OF FUSION, NUANCE INCLUDED

In spring 1951, the *New York Times* ran a sensational story: in Argentina, scientists had found the key to fusion energy.[7] By this point, fusion had been well-proven theoretically and there was a concerted effort to build hydrogen bombs with their uncontrolled fusion reactions.* In 1951, the quest for *controlled* fusion energy, though, hadn't really been formally contemplated in the US, UK and Soviet Union – the scientific superpowers at the time. The idea that scientists in Argentina – which at that point was largely a rural immigrant nation, with only about 16 million people living there – had achieved what those mighty powers hadn't yet really thought about, flabbergasted scientists the world over.

The story was based on a press conference held that March in Buenos Aires, where the Argentinian President Juan Perón promised a future where energy would be 'sold in half-litre bottles, like milk'.[8] Perón's 'new Argentina' vision meant playing in the same league as the US and Soviet Union, and it was this national push that allowed Austrian-born recent immigrant to Argentina Ronald Richter to convince the government to fund and build a 40-foot-high concrete bunker and fusion lab on the island Isla Huemul in Patagonia. The 'net positive result' Richter and his team reported in 1951 had been measured by a simple Geiger counter. The scientific community didn't take long to expose Richter as a fraud; and, for embarrassing the Argentinian president on an international scale, he was put in jail.

As much as 'Proyecto Huemul' and its *New York Times* coverage were ultimately laughed out of the fusion scene, they are often credited with being the trigger for serious controlled fusion research to begin. It turns out one Lyman Spitzer, a thirty-six-year-old astrophysicist working on the US hydrogen bomb programme at the time, received a phone call from his dad telling him about the *New York Times* article just before he left for a skiing holiday.[9] Aspen, Colorado, is well-known for its destination ski resorts; it's less known as the place Spitzer was inspired to conceive of magnetic fields being used to confine a hot plasma. With the notorious Richter announcement still fresh, Spitzer managed to attract the attention and funding of the US Atomic Energy Commission to pitch his idea for what became known as the 'stellarator'.[10] It's essentially a distorted

---

* Nuclear bombs use fission, while hydrogen bombs are based on fusion. They were known to be more powerful, hence the concerted effort to create them during the early period of the Cold War. The first demonstration of a hydrogen bomb occurred in November 1952, when 'Ivy Mike' was detonated in the Pacific Ocean, proving to be 450 times more powerful than the fission-powered Nagasaki bomb.

doughnut-shaped chamber surrounded seemingly randomly by magnetic coils around the ring. If you imagine melting a wedding ring and twisting and bending it in all directions, and then hooking many other misshapen smaller rings on to it, that's what a stellarator looks like – the misshapen-doughnut-melted-ring being the chamber in which the plasma gets confined; the smaller rings hooked on being the magnets used to do the confining. It's designed that way to try to herd the plasma into staying in the centre, away from the reactor walls. Spitzer convinced the US government it was worth pursuing and, only two years later, he was ready to begin experimentation.

Spitzer was unaware of the early UK research efforts focusing on a different approach, but James Tuck, who had previously worked with the British teams and was now at the Los Alamos National Laboratory, was inspired by Spitzer's funding success and applied for his own pot of money. His concept was named the 'Perhapsatron' – possibly the only instance of modesty in the history of fusion research – and his learnings from its development ultimately fed into the Zeta programme in the UK.

It's key to point out that Zeta wasn't a waste of time. Despite not show-casing breakeven capability, the experiments run on it contributed much to plasma studies in the years to come. The announcement was over-hyped, ill-timed and resulted in disappointment, but the science did continue to some extent in the background, before being shut down a few years later. The public relations damage meant the field slowed down and it suffered from that lack of trust, but the reality of the situation was simply that the science wasn't there yet, and could have got so much further, and faster, had that trust not taken such a hit in 1958.

The stellarator was to reign supreme in the fusion industry after the Zeta failure – that is, until the Russians unveiled what they'd been work-ing on all along.

In 1950, self-taught physicist Oleg Lavrentiev was serving in the mili-tary. He'd spent his tiny military allowance subscribing to academic jour-nals, and Moscow soon asked him to write a paper about the nuclear physics he was pondering, which he dutifully sent on. He wrote about controlled thermonuclear reactions, as well as his proposals for a hydro-gen bomb – which Moscow was likely most interested in, hence the call for the paper in the first place. His concept for the energy reactor was passed on to scientists Igor Tamm and Andrei Sakharov, who worked it into what would become known as the 'tokamak'.

In 1958, the USSR, the US and the UK met in Geneva at the Second United Nations International Conference on the Peaceful Uses of Atomic Energy, where the seeds of collaboration were sown with scientists meeting and discussing fusion research around the globe. This was the beginning of Cold War secrecy being lifted, meaning that international efforts could learn from one another and move fusion forward faster together.

In 1968, ten years later, the USSR announced that its scientists had managed to heat their plasma up to over 10 million °C and keep it there for ten- to twenty-thousands of a second.[11] It doesn't sound like much, but the stellarator hadn't reached over 1 million °C and confinement had never lasted any longer than a thousandth of a second. It was a UK team who were tasked with checking and ultimately confirming the result, to great celebration in the field.

Many planned devices that weren't tokamaks were abandoned in favour of this approach. Thus began the era of the tokamak.

Tokamak Time

The tokamak is a doughnut-shaped chamber that has multiple large magnets encasing the rim (tokamak comes from a Russian acronym that stands for 'toroidal chamber with magnetic coils').[12] What makes it different from the stellarator is that it isn't so misshapen, and therefore much easier to design and construct. Inside the doughnut, the gas of the hydrogen isotopes is heated up into plasma and the superstrong magnets are put to work confining the plasma to trigger fusion. The tokamak isn't only the most popular of the different magnetic confinement approaches, it's also the specific kind of reactor design that has been most employed worldwide since the Russian announcement of 1968.

Fast forward thirty years to Culham, UK, where a machine called JET (Joint European Torus) took the tokamak to the next level.

In 1997, the machine took an input of 24 MW and produced – using fusion – 16 MW.[13] It held the plasma at 100 million °C for two seconds.[14] Not breakeven, granted, but a hugely important display of our ability to create machines capable of fusion on Earth, and that ratio (denoted Q) of output to input of 0.67 still stands as the fusion reaction quotient record.

The demonstration was successful. Fusion was observed. The machine worked exactly as it was designed to.

And yet – the scepticism continued. Those in the industry knew that this demonstration was huge; it was what was needed to get those outside of the industry with the money and power to back the next stages of the technology's development. But still the general public sentiment balked at the slow-moving nature of the science, and the failings of the scientists to show that fusion was going to change the world of energy tomorrow. That 'fusion is fifty years away' headline came back into fashion, and again fusion was scoffed at.

It was almost as if the fusion industry had been crying wolf for years and was struggling to get the public's attention now the beast was pouncing.

But this wasn't the time for fretting about public sentiment. Those in the industry took the success of JET and used that proof point as evidence of fusion's potential to campaign for funding for the next stages of development: getting to that all-important breakeven point. They had to prove that fusion on Earth could produce more energy than was put in.

It was clear from JET's results that the route to breakeven lay in using a similar design but building a bigger machine.

The Curse of a Long-term Joke

Hype, it seems, was all that was needed to get the ball rolling, with Richter's *New York Times* headlines. But it was hype that also halted progress and ultimately had the public turn against fusion funding after the Zeta failure. In both instances, though, the hype was a form of ignorance, and instead of it being kept in check and put into context, it ultimately led to a decade of caution and scepticism. As a result, when the JET machine succeeded in showcasing the potential of the tokamak design, the excitement was hardly expressed in the mainstream press.

We're in another of the fusion hype cycles right now. Progress has been ramping up again over the last few years, and we're soon to unveil the results of that effort. If these soon-to-come announcements and media coverage are presented without context, there's a real danger that unmet expectations based on possible overhyped claims will land the fusion research field in another slump; maybe for another decade or, perhaps, forever.

JET's Big Brother: ITER

After the declassification of magnetic confinement fusion energy research during the Cold War, there had been talk in 1985 of a joint experiment where nations would collaborate and make fusion a reality once and for all. The US's Reagan and Russia's Gorbachev discussed the idea in Geneva; Europe and Japan joined in immediately at their request. The project was to be called ITER – the International Thermonuclear Experimental Reactor (later, the meaning was switched from the acronym to the Latin word 'iter': 'the way').

Then the bureaucracy began.[15]

In 1998, a decision was made to redesign the device to make it more affordable. The new design wasn't delivered until 2001. The US pulled out in 1999, causing huge delays, and didn't return until 2004. More nations then joined in, including China, Korea and India, which helped ease the burden of the multibillion-dollar costs that were racking up. But with the addition of more nations and so much bureaucratic history already wrapped up in the project, the central team became pretty sizeable, causing yet more diplomatic wrangling and delays.

After much discussion, a physical location was chosen in the south of France. The costs were to be split with Europe taking on about 45 per cent of the bill, with it being the host, and the rest split between the other nations. All nations were to play an active part in designing, manufacturing and installing the various pieces of the reactor, and all intellectual property was to be shared openly with all.

The agreement was signed in 2006.[16] The goal is to put 50 MW of energy in, producing 480 seconds of fusion, and to have 500 MW of energy come out.[17] Not only would this demonstrate breakeven, this would be ten times more energy out than in, a fusion energy gain of 10. The huge experiment would also aim to achieve a fusion energy gain of 5 but sustain it for longer periods of 1500 seconds.[18]

After the plans were agreed, it took until 2010 for ground to be broken and the build formally to start north of Marseilles.[19] By 2014, the costs had soared to ten times the original figure, at almost $50 billion, and the schedule had slipped eleven years.[20] At the time of writing, the first plasma demonstration is now scheduled for December 2025 and the first fusion reaction in 2035.[21]

When I first started reading about the recent history of fusion and ITER, I was shocked that I hadn't heard more about it in the media. When you consider how much coverage there was around CERN while it was being planned and built, it seems strange that an even bigger experiment, focused entirely on energy and curbing fossil fuels, was left in relative obscurity.

On one hand, the lack of public engagement over the years could be argued as a good thing: the scientists and bureaucrats were 'left to get on with it', as it were. On the other hand, with so much bureaucracy and delays and lack of public interest making nations question their investments in the first place, more engagement – possibly – could have held the project to account and perhaps it would not have been delayed for so long.

Hype can be a double-edged sword.

Making ITER Happen

ITER is set to be a 30 metre-tall device, weighing over 20,000 tons, with about a million parts making up the complex whole.[22] With the new 'turn on' date of December 2025 now set, there's good reason to be excited. Seven nations, Europe as a whole counting as one, are working together on arguably the most important problem of today: creating an alternative to fossil fuels. Other countries such as Australia, Kazakhstan and Iran look to be getting involved too. There is a level of optimism that both the plasma and fusion demonstrations will be successful, with the achievement of breakeven paving the way for fusion to make it into our global energy systems.

Unfortunately, it's not really that simple.

With any nascent technology that is still really at the science experiment stage, which fusion arguably is, there is much to be considered when it comes to spinning that science out of the laboratory and into the real world. To translate the ITER experiments into working power stations all over the globe, powering our planet with fusion, there are three technical issues that haven't yet been solved.

First, reactors that spin off from the work done at ITER need to be cheap, simple enough to build and easy to plug into the global energy grid. In short, they have to be accepted by the market.

The sad thing about trying to change the world of energy markets, as we explored in the batteries chapter, is that the health of the planet doesn't always take immediate priority when it comes to making decisions about our power. It comes down to cost. We might be able to create more energy than we put in with working fusion reactors of the future, but if we can't make it cheap enough to spin out commercially in the real world we can forget it. Endless energy isn't free when you consider the infrastructure that is needed to get it from A to B, and the costs of building and maintaining these fusion reactors.

In 1994, the Electric Power Research Institute in the US released a report titled 'Criteria for Practical Fusion Power Systems' to work out what exactly would be needed for fusion to be accepted in the market.[23] It found that fusion would have to be cheaper, easy to regulate and simply liked by the public.[24] In the same year, scientists found that the costs of building and running a tokamak-based fusion plant would be hugely more expensive than fission power plants, which we already have in the market, meaning an energy company faced with the choice would likely choose the latter, even if fusion was a working option.[25]

Second, the use of deuterium and tritium, the isotopes of hydrogen used as the fuel, makes maintenance of the reactor problematic.

When the energy is released from the fusion reaction, 80 per cent of the energy release comes in the form of so-called 'fast neutrons'. Neutrons, the name might suggest, are neutral – they aren't positively or negatively charged – meaning the magnetic field confining the plasma has no effect on them, and they whizz and collide with the reactor wall.[26] Over time, these collisions degrade the wall, meaning building a long-term power plant based on these reactions is problematic in terms of maintenance cost over the years. Instead of using both deuterium and tritium, we can use deuterium alone, which reduces the neutron output to only 35 per cent, but it is far more difficult to ignite and achieve fusion than with a mixture of the two.[27]

And it's not just about degrading walls. The bombardment also results in nuclear waste being created. The neutrons are travelling so fast that they knock atoms in the wall out of their structural positions and make them radioactive in the process, meaning radioactive substances will need to be removed and carefully disposed of.[28] The level of radioactivity in the waste created in fusion reactors is far less than in fission reactors, meaning the waste might not need to be buried quite so deeply to avoid contamination, but the volume of waste could be higher, and thus a problem in need of a solution still exists.

Third, we don't know if the fusion energy gain of 10 is going to be enough to achieve breakeven at the power-station level. When we take into account the energy consumption of magnet cooling systems and the coils inside the magnets that make them superconductors, as well as the efficiency of the plasma heating systems and other processes involved, ten times energy output might not be enough to leave us with a surplus. Some studies show that an efficient power station would require fifty times as much out than in, which means a different device altogether with a stronger magnetic field and more efficient physics, or simply one that is bigger.[29]

In short, ITER is very much an experiment, which if successful will bring huge knowledge gains to the fusion field, but translating those learnings into working power stations all over the globe cannot be assumed to be simple even if breakeven is achieved in 2035.

And with climate change already wreaking havoc on our planet, we might not have the time to wait for those technical challenges to be solved.

The Tortoise Might Just Be Too Slow

Technical challenges aside, ITER is still a hugely impressive project. But many questions throughout the years have surfaced about the way the project has been organised and run, especially as it is a government-run project as opposed to a corporate endeavour.

Setting up collaborations between international scientists, engineers, businesspeople, government officials and all sorts of support staff is a fiendishly difficult task. Finding common language and customs and cultures is not always quickly accomplished when starting from scratch with an enormous team.

And it's not just about getting people to play nicely together. The actual mechanics of the project have been set up not for efficiency gain, but for geopolitical compromise. There are over 1 million pieces, and ITER is being built from the bottom up, so every piece needs to arrive on time, in the right order. Many of the pieces are brand new designs of technology, and with invention comes delay. This is manageable in a small lab with everyone in the room working together on solving problems; it's a different story when those working on new innovations are spread across the globe.

All of the nations involved want to learn how each of the pieces of technology are designed and built, so that if ITER is successful, they can take that knowhow back to their own countries and kickstart their own fusion industry. It's not just that different components come from different countries; it's that different pieces inside every component come from different places as everyone plays a part in the build.

One example lies in the commissioning of the magnets. ITER will run on the biggest superconducting magnet system in the world. As you can imagine, getting to be the technology supplier to the biggest science experiment in the world is a hotly contested honour. But being the technology supplier to the invention which, if it proves to work using your design, will get rolled out to power the entire planet is an opportunity no business would turn down. The country in which that supplier resides stands to gain plentifully in tax, so, with seven entities working on ITER, it was decided that the fairest thing to do would be for multiple suppliers from different countries to work together in creating and manufacturing the magnets. They all had to work to the same design specifications, work remotely but in tandem, and produce in the same timelines two magnets each. International political compromise was prioritised over efficiency in creating the magnets, meaning that this one task alone was done at a much slower rate than if only one company had taken it on.

The upside to government contracts, though, is that employment of their citizens is important. The Italian magnet manufacturer is housed in what used to be a washing-machine factory, which was bought at a time when the company was failing. Experts were brought in to train up the washing-machine factory workers, and they became magnet engineers and got to keep their jobs.[30]

There's an argument to be made that global governmental collaboration is the best way to go for projects with such importance as inventing new energy sources. By putting the welfare of citizens and international fairness first, a better long-term solution can be created that benefits the many economically instead of the few. But time is running out for a viable solution to our reliance on fossil fuels, so maybe the corporate world, with its prioritisation of profit and, therefore, time to market, is better placed to getting fusion into the world.

WHEN SHARING ISN'T CARING

When someone invents something new, they can register that invention as their own intellectual property. They can apply to file a patent on their invention, so that it is protected from anyone copying their intellectual property and making money off the thing they created without their permission, or without fairly compensating them for their initial hard work. The wrangling of ownership and licences around new inventions is an area that can make or break new ideas in terms of getting them to market. For instance, if you invent something but don't own the registered patent or have an exclusive licence to use it, an investor is unlikely to want to give you money. They might worry that someone else – say, a bigger company with more experience and capability – might steal it and make a better business out of it, killing your company and their investment in the process.

So for ITER, with seven entities owning the project, one of which is Europe making the total number of nations far higher, and many suppliers being involved with many parts of the machine, crafting the intellectual property around the finished project has to make sense for all involved. What's been agreed is that all seven entities will get equal ownership of the intellectual property; they'll all get the final blueprints of all parts of ITER.

For some of the nations involved, the intellectual property will likely be made open IP in those countries – it will be put out, probably on the internet, for any company to take on and develop. This is great in terms of scientific knowledge; the knowhow behind making fusion happen (or whatever is achieved during the ITER years) will be made accessible for all and the world's most impressive feat, arguably, will be open for anyone to read and understand and revel in.

But the problem with open IP is that if everyone knows how to do it, what's the incentive for someone to invest heavily in taking those blueprints to market? Even if we know how to achieve fusion, energy companies, or whoever else it is that might take on the challenge of building the new national and global energy markets with fusion built in, will have to spend a lot of money working out how to scale the technology, run it in a profitable manner, train new staff and all manner of other activities ahead of actually making any money. If everyone has access, you run the risk of everyone waiting to be the second mover, so they can learn from the first mover's mistakes, and no one takes the initial bait. The energy companies won't know how quickly they will make back their investment in turning whatever comes out of ITER into market-ready power plants, so it may be wrong to assume that they would take on the challenge without owning the intellectual property in the first place to protect their investment.

In capitalist markets, businesses are predominantly market-driven. They do things because there's consumer demand; they can make money by selling whatever they're building to a customer of some kind. Yes, there are so-called corporate social responsibility and impact teams who take society's overall priorities into account, but this tends to be a very small portion of business focus.

In authoritarian countries such as China, businesses are market-driven to some extent, but the government can also mandate action from businesses under its rule. In other words, the government can tell businesses to act in a particular way, regardless of the market demands. In an event where China, as part of the collective of ITER, is granted ownership of the IP of a working fusion reactor, the government can order any or all of the energy companies within its jurisdiction to start developing scalable fusion power plants to serve the Chinese market, at will. This might mean that those companies won't make as much money for a while, and could very well result in poor outcomes for employees and their families if that results in layoffs, for example, but the businesses may have no choice but to comply with government rule. In the short term, this might not be great for the Chinese citizens associated with the energy companies but, in the long term, China might be able to speed ahead of other nations by simply moving first and making investments. If it manages to scale the tech into power plants that can serve huge portions of the country, and can make money while it is at it, there's a high chance that governments and companies in other countries – instead of doing the legwork themselves – would then license those developments from the Chinese companies, pay them to replicate the plants in other countries, or even simply have the Chinese companies supply the energy across the globe if it worked out economically viable to do so. We've seen a very similar story already with batteries and electric vehicles in China.

If this is the fastest route to fusion energy replacing fossil fuels around the globe, for some this might seem like not such a bad idea. You have the Chinese government mandating early investment, making it tough on its own countrymen and -women, to later reap the benefit of being the leader in the space.

But for others, the idea of having an authoritarian regime such as that of China controlling the world's energy is not so savoury, for much the same reasons already explored with batteries.

If the Chinese government doesn't mandate the scaling of the blueprints, though, and the markets don't show enough demand for the Western countries to build on the success of ITER, there is, of course, the risk that ITER is all for naught, and the world's biggest science experiment goes down in history as just that: an experiment.

'The Final Straw'

Consider for a moment that the ITER experiment fails in achieving the breakeven it seeks, though. The world's biggest science experiment, the most complex but important work being done by humankind to save our planet, the quest to prove what we know is possible and show the naysayers who, for decades, have called fusion a sham . . . doesn't work. How much more patience does the world have? Can society recover from yet another 'let down'? Of course, it wouldn't really be a let down; ITER is an experiment, and the work would still be getting us closer to the goal. But narratives are powerful and the ones around fusion are already fragile. The idea of ITER 'failing', in some sense, feels like it could be the last straw. ITER could be the solution to the world's energy crisis, or one of the most expensive failures in scientific history.

There's confidence that breakeven will be proven by the end of the ITER experiment. But that doesn't alleviate the concern I have around ITER disappointing the public. As more media coverage ramps up over the next few years as ITER's various milestone deadlines arrive, the public will surely also be 'let in' to the open secret that ITER is by no means perfect. With its technical and bureaucratic problems increasingly on show, there's a risk the public will lose trust in governments and scientists and a public backlash could quite easily be triggered from a few context-scarce headlines.

I worry that by writing about the woes of ITER here that the trust in fusion of you, the reader, will go down.

So with that, let me tell you why thinking of ITER as our only real shot at fusion is based on another false historical narrative we must set right, and why we must open our minds beyond it.

There's Gotta Be Another Way

With so much investment committed to ITER from the majority of the most scientifically advanced nations on Earth, there's a case to be made that ITER has taken money away from other credible efforts. With the commitment of 45 per cent of ITER's budget from Europe and about 9 per cent each from the rest of the nations, that's a lot of money out of federal alternative energy budgets worldwide focused on only one project. With federal alternative energy budgets continually being squeezed, there's not much left for other ideas after solar, wind, hydro, nuclear and ITER are accounted for.[31]

The idea that the industry is beholden to ITER, though, is a narrative that might not be so useful either. Some argue that, in fact, ITER is the source of problems for the industry, and the emphasis, and reliance to some degree, put on it by governments is what's holding back other methods from receiving alternative funding outside of government, to move them forward.

If we keep repeating and believing the sardonic quip 'fusion is always fifty years away', defeatist attitudes will take prominence. We don't have fifty years to wait for the first signs of fusion, and keeping on telling ourselves that it's either ITER or nothing at all blocks us from thinking more creatively and looking beyond those central governmental efforts to find out what else is out there. ITER's science might turn out to be the best, but solutions to its technical problems must also be solved in the meantime, ready for real-world application. If ITER's science isn't the best, we need to have other options to turn to, fast.

There are two ways of framing the tokamak focus. The first is that tokamaks are the best-studied way of doing fusion, ITER is doing good science with inefficient engineering, so let's have ITER move forward in getting the science right and, while they do so, build better engineering around it.

The second way, however, is based on the idea that tokamaks have disappointed us in the past, so why don't we just do things entirely differently?

There are other national governmental efforts in the fusion industry. For instance, in Germany, researchers have doubled down on Spitzer's stellarator design as opposed to the tokamak, with the Max Plank Institute's Wendelstein 7-X project.[32] With the stellarator design technically being far more stable than the tokamak, and with their 2016 first plasma demonstration (in which they managed to get the plasma to 80 million °C for a quarter of a second), there could be potential for this approach to make fusion a reality with their goal of confining the plasma for 30 minutes.[33] They're currently only at the same sort of stage tokamaks were with JET in 1997, though, but with their complex geometrical design, as opposed to the relatively simple tokamak dough-nut, the push towards a more continuous plasma confinement could prove an antidote to some of ITER's technical issues in making the fusion sustainable and cost effective. Japan has also got a stellarator device running in its LHD (Large Helical Device) project, the second biggest after Germany's efforts; there seems to be potential in this approach despite it still feeling quite early-stage.[34]

Beyond magnetic confinement, there are the national inertial confine-ment efforts; the biggest of which is the US NIF (National Ignition Facility), which focuses 192 high-power laser beams onto a small frozen hydrogen isotope pellet to cause a tiny explosion in which fusion happens, as well as the French Laser Mégajoule project.[35] These are all experimen-tal, just like ITER, and thus still lacking go-to-market plans that feel real-istic and doable. Also, both NIF and Laser Mégajoule are funded by the military, and given the history of nuclear physics and its ties to cata-strophic bombings, it's easy to see why these approaches are problematic beyond the early-stage explorative science.[36]

GOVERNMENT VERSUS CORPORATES

There's one more historical narrative around fusion that requires a little nuanced revising, and it's the idea that governments are better placed than companies at funding and leading development of early-stage complex science.

The question is: 'Are ITER or the rest of the government-funded experiments better for the world because they are state-run?' Governments have much larger budgets than companies, and even despite federal budgets for alternative energy being pinched, those budgets still beat corporate research and development investment by a long way off.

You could also argue that there's more accountability in govern-ment projects. There are more people to answer to if something goes wrong and it goes public, as it's an entire country relying on the insti-tution, as opposed to just private customers and shareholders. You could also argue that there's less chance of capitalism 'taking over'; meaning, you would hope, societal benefit would come before monetary profit, as the government is tasked with heightening the prosperity of the society over which it governs more so than simple economic gain.

But, on the other hand, we all know that politics tend to get in the way of fast process. Red tape, bureaucracy and geopolitical negotiation and compromise tend to slow things down. Averting the planet from climate disaster is a huge part of the rationale behind pushing fusion forward, so you could argue that government-led initiatives are the worst routes to go down, considering the fact that time is of the essence.

It is no surprise then that innovation is also happening commercially when governments can't be relied upon to move fusion forward beyond ITER, NIF or any of the other early-stage science experiments. Bringing fusion to the market, funded by an international consortium of governments, arguably would be preferable, with their bigger budgets and national accountability and social consciousness. But when all nations are insufficiently funding research across the board, specifically in energy as a whole, innovators intent on moving the world of fusion forward have no choice but to pursue their dreams elsewhere in the private sector.

More Baskets for More Eggs

There's a classification system called TRLs (Technology Readiness Levels) that is used to work out which stage of development new technologies are in. It was developed by NASA in the 1970s as a tool for the team working on a Jupiter lander to easily assess and discuss the various technology designs being proposed; it's now gone global and even been adopted by the European Commission as its formal system for assessing all science and technology projects applying for its funding. There are nine levels starting with TRL 1 and 2, which refer to basic research, up to TRL 5 and 6 where the tech is being demonstrated outside the lab, and TRL 8 and 9 where the tech is launched and real-world operations are being proven.

The idea behind the TRL system is to allow decision-makers such as investors, policy-makers and other people with skin in the game to have an understanding of what stage a new technology is at, without having to know the full industry and science behind it inside out. It's a helpful way of fast-tracking understanding of context when it comes to researchers or science and tech companies applying for grants, which can be a huge burden for innovators if they have to fill in forms and explain complex ideas every time they want access to public funds for research and development.

When it comes to trying to solving big, complex problems and answer impactful questions such as 'What is an energy source that can replace fossil fuels that can be safe, clean, unlimited and cheap enough to replace what we have today?', it makes sense to hedge your bets across many different approaches to ensure the solution is eventually found. For example, NASA will spend about a third of an initial project budget on projects in the TRL 1–5 range, and the rest on projects mixing technology across TRL 2–7.[37] This is to avoid putting all its faith in one new technology that might end up not working. It doesn't want to have to keep going back to the very beginning, so by taking on a few different approaches at different levels of maturity, it covers the risk that comes with trying to do big, complex things.

Within the fusion industry, there's an underlying feeling that we've put all our eggs in one basket. We've picked the tokamak as the main technology of choice outside military applications, and pointed almost all our public funding globally into one big concerted effort to make it work, in the south of France.

ITER, over the decades, has sucked a lot of money out of the rest of the fusion industry with its high-level political standing, international reputation and mental capture over the supporting industries. It seems a lot of people have come to believe that it's the logical project to focus everything on, considering how much we've already spent on it.

But the sunk-cost fallacy is exactly that, too: a fallacy. There's inertia in terms of trying new things. The idea of 'starting again' feels so enormous, not many are willing to dive in with their time or their money.

START UP, NOT START AGAIN

There are people who believe passionately in fusion's potential, but see another way beyond the governmental efforts. They are the fusion startups.

A wave of small private companies popped up in the late 1990s, with TAE Technologies kicking off in 1998, focusing on a different magnetic confinement design. It combines elements of many other designs throughout fusion's history, called a 'field-reversed configuration', which is simpler in design than the tokamak.[38] The company also focuses on the isotope boron-11 and protons themselves as fuel, as opposed to the isotope mix of deuterium and tritium, reducing the radioactivity of the output. In order to get this design of machine with that kind of fuel to confine the plasma and trigger ignition, plasma needs to be heated to billions of degrees, as opposed to the tokamak's millions, and the boron–proton fusion reaction doesn't give off as much energy as deuterium–tritium. The company had been notoriously secretive until it started publishing more of its efforts in scientific journals, speaking to the press and creating its first website around 2015. Still, the technical challenge plus the secretive approach haven't stopped investors such as Microsoft co-founder Paul Allen, Goldman Sachs and the Rockefeller's venture-capital firm Venrock putting in approximately $800 million so far.[39]

TAE Technologies' approach, though, might be just what the investors are buying into. The company, like most of the fusion startups, seems to put engineering first and science second, meaning, it is not waiting on the results of ITER or the other science experiments to confirm that huge unwieldy machines can do fusion: it is refining and testing engineering solutions to see if it can get it to work in other ways. As such, the researchers are starting with the deuterium–tritium mix as the fuel as they build their first prototypes, and will work up to proton–boron. They know that to ultimately get fusion to the masses, they need to think commercially about the market and realistically in terms of scaling up; not just prove they can do it.

In 2001, another company with an alternative approach to fusion was created by Québécois plasma physicist Michel Laberge: General Fusion.

On his fortieth birthday, Laberge decided to quit his job at a laser-printing company and focus on solving the energy crisis by building a working fusion reactor. He decided to dust off an old 1970s reactor design from the US Naval Research Laboratory, and pair that with modern computational methods, called 'magnetised target fusion'.[40] Jeff Bezos, founder of Amazon, is one of the key investors in General Fusion, and along with others such as the Canadian oil giant Cenovus and the Malaysian government's investment arm, a tidy $81 million has gone into building General Fusion's machine.

The machine, which has a steampunk feel to it, consists of a metal sphere surrounded by fourteen cylindrical canisters poking out at all angles. If you imagine one of the little spiky pieces in the game of jacks, but blown up to about 5 metres tall, that's what the machine looks like. This is just the initial version; the final version the company is aiming for will have a forest of around five hundred canisters protruding out. Inside the sphere is liquid metal, which is fed into it in such a way that it swirls, creating a centrifugal force that keeps it held to the inside wall of the sphere with an empty centre to house the plasma. The canisters hold pistons inside, and they all fire at the same time, and acoustic waves squash the liquid towards the plasma, pressurising it and hopefully heating it to the 150 million °C to get the nuclei to fuse. The benefits of the General Fusion approach is that the engineering is well established and the parts are not too expensive, so building more reactors is realistic. The liquid also doesn't degrade because of the fast neutrons emitted from the deuterium–tritium fusion, which means upkeep is easier. Without advanced computation, though, it's hard to synchronise the pistons effectively and therefore the plasma is tricky to confine just right, and so work still needs to be done.

One of the newest startups to enter the race kicked off as a result of the US Department of Energy eliminating funding for an experimental fusion reactor at MIT.[41] The leader of the lab, Dennis Whyte, was dismayed at the decision, and soon decided with his colleagues and graduate students to continue the work as a company instead of a research lab. Commonwealth Fusion Systems (CFS) was born, and it promptly attracted the interest and investment from both an Italian oil company and Breakthrough Energy Ventures, a fund led by billionaires Bill Gates, Jack Ma, Richard Branson and – again – Jeff Bezos, to name but a few.

CFS has doubled down on the tokamak, but it is keen to get around some of the economic issues in translating the science into the market by building a much smaller, cheaper and modular version. For example, the magnets confining the plasma utilise a superconducting tape, which makes them more resilient, far more powerful and a lot smaller than their rigid counterparts at ITER. CFS also turned the problem of the radioactive deuterium–tritium fuel into an advantage by building lithium salt into the walls of the reactor to collect the neutrons, which then creates more tritium in the process. Deuterium is easily sourced from seawater, tritium has to be made, so having this circular element built into the machine itself is incredibly clever. (General Fusion also does this through using its liquid metal in the sphere.)

All the startups are aiming to have economically viable power plants in the 2030s by demonstrating breakeven around 2025. There are plenty more startups, and really it's anyone's game when you consider the fact it's not really about having the best science but more about being able to create something that the market will actually accept. We only need one startup to demonstrate that its reactor is able to achieve that all-important breakeven point, and get a net-positive energy gain as a result. With a cheaper version proven by a startup, it's highly likely that an energy company would buy the startup and then use its own power-plant-building knowhow to scale that up and supply the worldwide energy markets. We don't know yet which one of the reactor designs will succeed first, hence why so many startups are throwing their hats in the ring in the quest for such a huge prize, but out of around twenty fusion startups, only three are focusing on the same technology as the billion-dollar, slow-moving ITER.[42]

RISKING THE REWARD

Investors tend to hedge bets on companies, not technologies themselves, in terms of spreading their risk. They'll decide that a particular innovation – such as self-driving cars – is one to bet on, and then they'll spread their risk across the many companies and innovators pursuing that advancement. To get a venture capital firm to decide that a particular industry is worth investing in takes a huge amount of time, as they'll want to be sure of their conviction.

Venture capital firms (the companies and people within collectively called 'VCs') also care about outside perception of their investment moves. They have their own funders, who trust them to put it into sensible companies and schemes to provide them with a return, who the VCs have to keep onside. They also want to be seen as innovative and fast-moving on new opportunities so that more people want to get involved with investing in their fund, and so the best companies agree to take their investment and work with them. Startups won't just take money from any VC. For the trendiest companies looking for money, VCs normally have to compete with one another to get the startups to say yes to them. If they don't want to compromise on the amount of equity the startup is giving to them, they have to prove themselves as the most promising and useful investor for the startup to agree to beneficial terms. Think of *Dragons' Den*, and how the personality and experience of each of the dragons is brought to the table when they are competing against each other to get involved in a good company; it's the same for venture capital firms. A VC needs to be seen as making the right bets, and being the firm to have taken the right big risks and benefitted handsomely when everyone else was too scared to get involved.

Getting VCs to take a firm stance on a prediction that a particular high-risk piece of technology such as fusion will revolutionise the world and provide huge profit in the process is therefore no simple task.

At the time of writing, it's predominantly wealthy individuals who are putting money into fusion; some because they really do believe in the companies and their ability to get to a working fusion power plant in the next few years, others because they believe fusion must be funded if we're to save our planet. Many are 'taking a punt', as it were. The investors jumping in are interested in the 'moonshots' of today, taking huge risks hoping that a huge reward will come eventually, as they can afford to do so.

The problem with investor sentiment around fusion more generally is that it still seems too risky for more mainstream VCs to consider. Time is crucial when considering an investment; the time it would take for the

VCs to be paid back handsomely is usually expected to be around ten years at most. This simply isn't enough time for fusion, especially if you don't have complete trust in the startups gunning for the 2030s. That insecurity is too much for VCs to bet on without angering their funders and risking their reputation in the process. Some VCs look at those funding fusion and applaud their courage, but view it more as a billionaire almost donating their money to science, as opposed to making a smart investment.

It could take billions of dollars, not millions, to get to the point of commercialisation, but if enough wealthy individuals as well as VCs put in millions, it adds up, and the prospect of fusion becoming a reality starts to feel more realistic.

The narrative that fusion will always be fifty years away doesn't help instil confidence in the investment community. The narrative that ITER is the best bet doesn't encourage VCs to look to the startups who have joined the race.

The reality of getting more investors interested in fusion looks like this: the first stage is finding the investors interested in businesses that are working in the climate-change sphere. Then within those, you need to find the ones who believe there's a market for disruptive solutions in the energy field, as opposed to the already established alternatives. Within those, you need to find the ones who can get past the hype in the idea that renewables alone can work, and can see that more needs to be done. Further shaving off the pool of investors means finding those who believe that traditional nuclear is not the only option beyond renewables, and finally within this already pretty tiny segment of investors who have enough money to really make a difference to a company operating in the space, you have to find the one that says: 'All right, let's give fusion a go.'

If startups are to be the ones to revolutionise the field and speed us further towards a fusion future, narratives must convince investors that putting money into the field is sensible economically, and will make them look good among their peers and their own funders. Society cannot afford for historical narratives lacking nuance to prevent those with the capital available to take the leap in funding the most promising technology in ending our reliance on fossil fuels.

Is It Safe?

It's not just the investors funding the industry, and the energy companies expected to turn scientific proof into working fusion power plants for all, who need to be brought on board with fusion. We the public also need to trust and believe in fusion as the best option, and there's one more narrative that blocks support for fusion: the idea that developing fusion energy is not safe.

Chernobyl was a horrific nuclear disaster. When safety tests weren't properly carried out in the Ukrainian plant in 1986, the disabling of many emergency safety systems resulted in an uncontrolled nuclear chain reaction, which created a huge steam explosion. For days, the radioactive plumes from the fission reactions wafted into the surrounding atmosphere, resulting in a widespread evacuation of the surrounding area, which is still unoccupied. The total number of deaths as a result of the radiation is still a matter of disagreement among different bodies; the UN estimates 4,000; Greenpeace estimates 20,000.[43]

What we can be sure of, though, is that Chernobyl and subsequently the nuclear disaster in Japan's Fukushima (triggered predominantly by a tsunami in 2011) have reinforced the unsafe narrative around fission nuclear energy in the eyes of the general public. The words 'nuclear' and 'safety' simply don't go together.

But nuclear fusion and nuclear fission are not the same thing. Yes, there's the low-level radiation created by the deuterium–tritium reaction within the reactor, but the set-up is such that an uncontrolled chain reaction releasing radioactive fumes into the surrounding area cannot happen.

The other side of the safety coin is the thing society seems to always be most fearful of: nuclear war. Further research and development of fusion energy reactors might lead to better understanding of how to improve the performance of hydrogen bombs.[44]

In one sense, this is very worrying, particularly considering two of the big governmental efforts are funded by the military: NIF in the US and Laser Mégajoule in France. However, governments are already in possession of catastrophic weapons, our atomic bombs (fission) and our H-bombs (fusion). We already have unmanned drones. We already have tense geopolitics. We already have huge armies the world over.

We already have the ability to cause huge damage; halting energy research in a bid to stop the creation of even bigger bombs isn't going to stop wars.

No technology is flawless, no technology comes without downsides; and anyone telling you otherwise is serving you hype. Saying that, we must be able to work out which of these pros and cons are rooted in reality and which risk profiles are being represented disproportionally. We must be able to work out which narratives are simply rooted in historical clichés, without the associated nuance. We must be able to work out which narratives are unfairly trapping society in fear, misunderstanding and simply out-of-date ideas, making us unable to take in anything new that may go against firmly held views.

We also must be able to separate Chernobyl and H-bombs not only from fusion but even from the idea of unsafe technology. All technology is unsafe in the hands of the wrong people; tools can be wielded for both good and evil.

TRUTH AT THE EXPENSE OF TRUST?

I'm a fusion optimist and I really want this technology to exist in the world. I'm also very aware of the fragile nature of the narratives around it, and how easily the spell can be broken when negative stories about fusion's development capture the public imagination.

Should journalists and science writers pull back from covering fusion fully out of fear of making people less optimistic about its progress, even if they know there's overhyping at play? Or should they keep up the enthusiasm in some kind of cheerleading effort to ensure we get fusion over the line without the worry of bad public sentiment? Thinking about the fusion industry as a whole, warts and all, is not about being a naysayer, but rather a realist.

Knowing our human tendency to polarise and have a definite 'answer' around complex topics, there's a responsibility to not fuel the fire of either the 'fusion is definitely coming soon and it's awesome' or 'fusion will always be at least fifty years away and it's not the best thing to do anyway' sides of the debate. If I had to choose one 'side' for the public to take, I'd want people to be optimistic, regardless of the ups and downs of the industry, as that positive sentiment is needed to keep the funding and focus going. If a fusion-powered world is to become a reality, public optimism is needed to get there.

It's hard to critique something without prompting people to lose overall trust. The best way to keep that trust is to explain as fully and thoroughly as possible, which is not always doable in a short article or a tweet.

The fact is, though, there's reason to be both optimistic and pessimistic about fusion, and neither is fully correct or incorrect. We have to trust that we as a society can hold conflicting narratives in our heads at the same time, and not default to damaging historical absolutist reductive narratives.

We all need to look at our own ways of looking at the world, and reconcile our human tendencies to look for definitive answers at the expense of a maybe harder to swallow nuanced truth.

If we decide that the world needs fusion, then fusion, in its imperfect state, needs our trust.

Removing the Space-based Rose-tinted Glasses

It's been fifty-one years since man first walked on the moon. Fifty-one years since Armstrong and Aldrin left their dusty footprints on our nearest celestial neighbour in the name of humankind. Fifty-one years since a huge government financial intervention made real the dreams of scientists, according to Richard Nixon, in the 'spirit of brotherhood, a spirit of the fellowship of human achievement'.

But it wasn't goodwill that incentivised the US government to plunge billions of dollars into the NASA programme – it was the prospect of winning the space race, trumping Russia, and massaging the ego for short-term political profit.

Of course, the moon landings were an extraordinary adventure. Never before had humankind ventured so far, and the ability to broadcast it live meant the whole world experienced it from their living rooms. The future felt so close, anything felt possible, and technology felt like a limitless medium to transport us all to a better world. 'One small step for man, one giant leap for mankind' are remembered as the words at the root of one of the biggest feats in human history – to be remembered by all as impressive, momentous and right.

But with the motivations for heading to the moon in the 1960s lying firmly in Cold War politics, was the 'one giant leap for mankind' sentiment almost an afterthought? We're not any closer to global 'brotherhood' and 'fellowship' than we were in 1969. Is this really something that the moon landings could have brought us? The fact that funding for the missions was cancelled not long after Apollo 11 suggests that furthering science and forging international collaborations towards a better world wasn't really the focus of those who had the power to make it happen.

In fact, if you put aside the Nixon and Neil Armstrong quotes and instead look at what the US Information Agency said in 1960, you'll find the focus on space as a tool for power wasn't so subtle: 'Our space program may be considered as a measure of our vitality and our ability to compete [against] a formidable rival, and as a criterion of our ability to maintain technological eminence worthy of emulation by other peoples.'[1]

Fifty-one years later, the space race feels like it has been revitalised. The modern-day industrialists are wielding their financial power while the mature corporate space industry fights to stay relevant. Jeff Bezos, the founder of Amazon, founded the rocket-building company Blue Origin in 2000. Elon Musk, the founder of X.com (which went on to become PayPal and be sold to eBay for $1.5 billion[2]), founded the aerospace company SpaceX in 2002. New countries are entering the publicly funded outer-space endeavour alongside long-standing governmental efforts elsewhere – since 2010, we've seen Australia, Portugal, Belarus, Mexico, Bahrain, UK, Bolivia, New Zealand, North Korea, Paraguay, Turkey, Poland, South Africa, Turkmenistan and the UAE all create their own national space agencies. It's safe to say that space business is not a futuristic concept, but an already burgeoning industry.

The question is: what's the mission this time round? Global fellowship isn't really touted as the dream any more when it comes to making a market out of space. Despite Carl Sagan's famous vision of 'the human future in space', is there any evidence that this time things will be different, that the mission is that of human exploration, expansion and curiosity, as opposed to that of corporate power?[3]

The marketing materials of modern space efforts, whether commercial or governmental, tend to show one of two things. Either a vision of a peaceful, thriving society – a utopian community where everyone gets along and each human has realised their potential. Or a vision of ambition, entrepreneurialism and the intrepid inventor-explorer building a better world. In short, they show something quite unlike the reality of Earth-based society – where, yes, we've advanced much in the fields of medicine and technology, but we still have war and inequality. There's no mention of when this shift will happen – when society will 'catch up' or reinvent itself; the picture painted assumes society will have 'sorted itself out' before, or as a result of, access to space.

This sentiment – 'space will be different' – is a compelling one, but it may also be the idealistic blinker blinding our peripheral vision to everything else that is going on in the bid to commercialise beyond the Earth. Removing that blinker means seeing the space commercialisation industry beyond the hype: an industry much like any other we've had throughout history. One that is impressive, visionary in ambition, far from perfect and, ultimately, with inherent power structures no different from what we've seen before.

Assuming the space industry will magically produce a different, better society than the one we're currently in – based on the way these efforts are presented – means we cannot spot when things go south. Societies in the past believed those idealistic images of the oil industry ('power for all!'), the colonial explorers ('whole new worlds!') and the pharmaceutical industry ('everyone can be healthy!'), only to realise the inherent problems only once the power accrued was too large to contend with.

If we blindly believe that the space industry will be different from those that have come before it, we run the risk of repeating the mistakes of the past and missing out on the potential that space really *does* afford us.

There's a romanticism about space, and understandably so. The dream of humans graduating beyond Earth to explore our wider solar system, with some of the most incredible technologies we've created, is a compelling one. And the opportunities in space for better mapping, understanding and connecting of our world contribute to understanding climate change, internet accessibility and global logistics efforts. Having romantic notions about space isn't wrong, but that is only one side of the story, and we must all consider our romanticism alongside the reality of space as an industry, and be careful not to be swept up and blinded by our love.

THE BUSINESS OF SPACE

In 1984, Ronald Regan signed into law the Commercial Space Launch Act.[4] It essentially said that NASA had to look for commercial options for the technology utilised in its missions, if it was indeed readily available on the open market. The idea was to both stimulate the commercialisation of space technology (and hence ensure the country was making back some of the science investment in corporation tax), as well as save money on 'in house' development by instead contracting out to those who could do it better and cheaper.

Since then, the commercial space industry has been growing steadily, but the last few years have seen a huge growth in interest, funding and the number of companies in operation.

In years gone by, it was only government that could overcome the upfront cost of access to space. Nowadays, the modern-day Rockefellers are the internet-era billionaires such as Richard Branson, Elon Musk and Jeff Bezos, and they are investing huge amounts of their wealth in the space industry, alongside the venture capitalists coming in and providing private investment.

So what's changed? For starters, it's simply easier to start a space company now. Splitting the industry up into a few buckets: there are the rocket builders, the rocket launchers, the satellite operators, the telecommunication and tracking companies, the data analysis companies, the exploration companies, the research companies – the list goes on. Enabling technology is smaller, cheaper and exists on a scale like never before; there are more investors interested and hence more upfront money available before revenue is made; and with different businesses covering different parts of the process, it means companies can in some sense 'slot in' to the industry much more easily by making the most of the services already on offer and build on top.

A note before we move on – the growth of private companies in the space sector is broadly a good thing. If we agree that getting better satellites into orbit, and providing more internet around the globe, and exploring other worlds such as Mars and the moon, and doing experiments in the International Space Station are all beneficial to humanity (which I certainly do believe), then we want the best technology used and the least amount of government spend wasted in the bid to make all that happen. Competition (which drives down prices and drives innovation and quality up) is impossible in a government-monopolised industry, and is what commercialisation can sometimes very positively bring.

Wanting clarity about what's going on in a field is not the same as thinking the field is universally bad. Without clarity, though, it's easy to miss the not-so-savoury elements that can't be found in the hype.

What It's Not

Just because there's lots of activity in the space industry doesn't mean that what's going on in the modern-day space race is clear to those on the outside. There's plenty of coverage in the mainstream press of the two visions of space – the more utopian 'sci-fi-becomes-reality' stories surrounding space tourism or moon villages, as well as the 'cult-of-the-entrepreneur' stories surrounding Elon Musk and Jeff Bezos. There's little coverage beyond. If you were to therefore take the temperature of the general public's attitude towards what's going on in space, it's unlikely it would match reality.

The media hype about the plight of the entrepreneurs means that we hear plenty about the delays and the failures of the plucky startups trying to get humans into space, but little about the bulk of their day-to-day business, and how they make money.

We hear about Virgin Galactic seemingly always being behind schedule in its bid to provide space sightseeing flights.[5] In July 2008, they were meant to be in space in eighteen months, but the same was true in December 2009 and April 2011. In May 2013, Richard Branson said he'd be on board their space flight on Christmas Day that year – needless to say there was no such flight. A few more 'in a few months'-type comments and, in 2014, the *Sunday Times* even wrote a piece titled, 'The $80m Virginauts stranded on Earth', echoing the disgruntlement of those who had prepaid between $200,000 and $250,000 to get a spot above the skies.[6] In 2018, Virgin Galactic finally announced its first space flight – though it was still a test flight (and so had none of the over 700 already-paying customers on board) and, in reaching 80 km above Earth, it didn't even get to what's formally defined as space by many international space agencies (100 km).[7]

At least Virgin Galactic is still operating – unlike XCOR, another space-tourism company that took $100,000 each from 282 customers before going bankrupt in 2017, unable to pay any of them back, never mind take them to space.[8] Again this story played out in the mainstream media, which interviewed the disgruntled customers and showcased space tourism as still so far away.

These high-profile difficulties and failures playing out in international media can give the impression that building any kind of business in space is prohibitively hard. It makes it look like those working in this area are so far off success it doesn't need proper consideration and questioning, beyond surface-level taunting of the failure of the companies, their founders and the arguably daft rich people who paid them money.

Then there's the coverage of the outlandish ideas – the ideas which, in time, may prove to be what's really in store for humanity, but for now are proving to be the dreams of the perceived 'crackpot inventors'. The fact that space is so closely linked to sci-fi sometimes makes it difficult to tell whether what is being proposed is fiction or genuine mission, but these are the stories the public like to read and thus the media will deliver.

Take Mars One, for example. Launched in 2012 with its proposal to send people on one-way missions to Mars, it was to be funded by turning the whole thing into a reality TV show – the website even touted the income from Olympics broadcasts as supposed proof of its ability to raise enough capital from broadcasting to create a society of humans on another planet. Mars One 'recruited' 100 people to live the rest of their lives on Mars (I say 'recruited' as it was essentially an online popularity contest with contestants even paying their way up the levels of the game), only to go bankrupt in January 2019.[9]

And these outlandish ideas, understandably, get coverage – Mars One was all over the media throughout those years – but its hype can also make those following the industry from afar sneer at the idea of building a space business; thinking that all efforts are just as far off and 'crazy' as that of Mars One and Virgin Galactic. This kind of narrative only leads people to treat 'keeping up with industry' as entertainment, instead of engaging with it as they might with other kinds of current affairs.

Most of the ideas that gain media attention centre on the projects that give those following them hope that they might, one day, get to look down at the Earth from above. Human spaceflight ideas are not unfathomable, but they feel far off, and the 'it's so far away' sentiment sometimes represents the extent to which the general public engages with the industry. And when human spaceflight efforts do come into fruition – which may be shortly as space tourism is now gaining much momentum – the excitement understandably will be huge, but that media focus will still be on such a small sliver of the space industry. This is all despite the reality that the space industry is, currently, ever growing in sometimes problematic 'behind-the-scenes' power.

Just like the pursuit of even better batteries for our smartphones takes up the attention of more Western consumers while the reality of what's going on in the cobalt and lithium mines is hidden away behind the scenes, so too are we distracted by the consumer space-travel narrative when it comes to considering the future of space. We're not all really seeing what's going on in the industrial fields already making huge gains above our heads. We're really not grasping what the world of space commercialisation actually looks like.

All of the fast-growing industries of years gone by, such as mining, energy or healthcare, had (and continue to have) both huge benefits for society as a whole and deep-rooted problems for humanity worldwide.

The space industry is no different.

The Already Booming Space Business

I previously mentioned the 1984 Commercial Space Launch Act, but the first real piece of regulation in space commercialisation came even earlier – in 1962 when John F. Kennedy signed the Communications Satellite Act, paving the way for private companies to own and operate their own satellites.[10]

We often forget about satellites when space commercialisation is discussed, despite their influence on our daily lives. They seem like old-school space technology: indeed, the first commercially sponsored satellite was launched in 1962, funded by Bell Labs and AT&T, and allowed for live television broadcast between the US and Europe, only five years after Sputnik.[11]

With the growth of the internet, and as we push towards getting the 'other 3 billion' (the term often used for those without internet globally) online, as well as the rise of the 'internet connected' device both in homes and in industry (think Amazon's Alexa, as well as factories running autonomously), the need for more, better satellites providing stronger connection has grown exponentially. This has pushed the satellite industry beyond GPS, television broadcasts, mobile phone connections, weather forecasting and global imaging, into the realm of ever better internet, higher quality imaging, an ever-growing mass of data, and beyond.

Indeed, the companies such as SpaceX and Blue Origin that get the mainstream coverage – for ideas such as 'colonising Mars' and space tourism – are actually making their money launching government and business satellites into orbit. Instead of the general discourse focusing on the real bulk of the space commercialisation industry, we're caught up talking about what a human habitat might look like on the red planet.

The approach that many new companies are taking is to create constellations of small cheap satellites – instead of launching one or two huge expensive ones – which, due to this constellation network effect, can provide coverage across huge areas. This is the approach of companies such as SpaceX and even Amazon, as well as space startup OneWeb, arguably the front-runner, and it's trendy because it means they can offer high-speed internet at a much lower cost than the existing players – in a market that is not short of customers.[12]

In the past, there has been a huge bottleneck in getting satellites into space because of the lack of capacity of the launch industry: there just haven't been enough rockets being launched that satellites could be boarded on. If you wanted to get your small satellite into orbit, you'd have to piggy-back on a commercial rocket – maybe a SpaceX one – if it happens to have excess capacity that it didn't manage to sell to a higher bidder (just like hotels have cheap last-minute spare rooms). The small satellite, then, hitches a ride to the orbit that the main payload was built to go to. Nowadays, there are many rocket-launching startups, such as Rocket Lab and Vector, that focus specifically on the small-satellite market of nano- and micro-sats, and launch them specifically into the orbit *they* require.

Now, space is big, but when it comes to satellites, there are particular orbits and altitudes that make the most sense for satellite businesses. The prime locations for Earth observation satellites (such as those focused on weather or surveillance) are concentrated in the heavily used polar orbits – as they allow for satellites to be close to Earth and see almost everything on the planet in only ninety minutes. This means many orbits overlap above the North and South Poles. With more and more satellites vying for the same altitudes, as well as more opportunities for satellites to arrive above the North or the South Pole at the same time, the chance that a collision will happen is growing.

When you consider what the role of each satellite is – maybe one is providing internet access to large swathes of Colombia, or maybe another is providing GPS for airplanes flying over Sweden – the size of the collision problem (and possible destruction of the corresponding satellites) becomes quite clear.

And it's not just about the new satellites being launched: we already have loads of space junk in orbit and there are plenty of collisions that have happened, preventing the operation of expensive, useful satellites.[13]

Then there are the decommissioned or dead satellites from years gone by, which we've not invested in returning to Earth (so they are essentially just bits of metal going around the planet), as well as debris from previous collisions (you might not just get a dent in your satellite if there's a collision, the two might break into multiple pieces, and then some of those bits might continue to orbit at speed, posing a threat to other satellites, while the other bits come out of orbit and burn up in re-entry to our atmosphere). This problem is only going to increase with more satellites and more collisions in the future.

Many of the new satellites that are being launched are nano- or micro-sats. If they are orbiting at lower altitudes, the Earth's atmosphere creates enough drag that, over a couple of years, they will burn up and decay – thus not posing as much of a debris problem once they are finished with their mission. This is not so true even for small satellites at higher altitudes. There isn't as much atmosphere, and hence there is less drag pulling them down to burn up and decay. These satellites, once their mission is complete, could end up staying in orbit for tens or hundreds of years, despite being defunct. These small satellites – although better in terms of space junk sustainability than the large satellites of years gone by – still orbit the Earth at great speed, and thus present a problem when it comes to collisions.

Cleaning Up Space

Space debris is not a newly discovered, or newly dangerous, environmental issue, but with sustainability of the Earth becoming a more mainstream topic – particularly with regard to plastic in the oceans – the space-debris debate has recently become much louder, with more solutions and preventative measures being proposed each year.[14] There have been ideas surrounding spacecraft with harpoons, robot arms and attached nets flying around gathering up the trash, as well as companies that attach magnets to new satellites so that they can easily be brought back to Earth once their life is up, as opposed to leaving them up there.[15] But most of these efforts are in their infancy, one-off missions or simply ideas.

Space debris is a problem, the problem is growing, and there are solutions out there that – if implemented – work. It's beneficial for everyone if the problem is solved. Go to any space conference and people are talking about it with appropriate concern ... and yet, there is no international solution to sort out the space debris problem.

Just like any other sector that is commercialised to some degree, money comes into play. Space is no different.

So, we have companies proposing new technology and ideas for solving the space debris problem, but they aren't getting off the ground. For these kinds of sustainability space companies, the business models are tough to get right: getting satellite companies to pay to have their tech removed once dead or damaged, or pay to have tech installed before launch to ease the de-orbiting process, is a tough sell. This is because these space-debris mitigation efforts involve an upfront cost to the satellite company. Leaving satellites up in space after they are dead or damaged doesn't cost anything, so asking the companies to be responsible is more of a long-term, socially minded play, not a short-term, business-profit one.

These kind of mitigation efforts only work if everyone is on board – but once a few companies buy in (frankly, out of the goodness of their hearts), fewer companies will be incentivised to join them: the problem will be lessened by those who buy in, so those who don't will be somewhat 'off the hook'.

Trying to get all the satellite companies on board – from the traditional industry stalwarts to the young startups doing business with SpaceX – is almost out of the question without the regulators getting involved and essentially forcing the matter. But getting one country's regulator to agree to putting in place rules demanding that satellite companies pay for the clean-up of space is also tough. If one country puts in space-debris mitigation rules, such as demanding satellite companies install magnets on board before they launch, the companies will simply relocate to countries where this kind of costly prerequisite isn't demanded. Countries are incentivised, therefore, not to put regulation in place as they can earn tax from the disgruntled companies that relocate to their shores.

This isn't just theoretical either – a US company Swarm was denied regulatory approval to launch its experimental 'pico-satellites' at home in the States (due to their being too small to be tracked in orbit, interestingly), so it instead moved its launch to India and took off from there. The US Federal Communications Commission did investigate this blatant avoidance of safety regulation, and fined the startup $900,000.[16]

For the space industry, that relatively low price-tag won't really deter future bad behaviour of that ilk, and not everyone will be caught or even pursued. In fact, only one year after its fine, Swarm raised $25 million in venture capital funding, so it clearly hasn't deterred investors.[17]

COMMON TRAGEDIES

The 'tragedy of the commons' is an economics concept referring to a situation where there are multiple people involved in one system with shared resources, all of whom act independently according to their own self-interest. In the process of doing so, as a collective, they run down or spoil said shared resources, so much so that the whole thing is ruined for everyone.

You can see it happen in the case of the climate crisis we're all currently in, or in the case of overfishing the oceans. Individual countries, companies or people themselves act in their own best interests, and by collectively doing so ruin it for everyone.

This tragedy of the commons when it comes to regulation in space is a well-known issue within the industry, and the problem is no different in space than in any other facet of society. And just because an issue is widely recognised doesn't mean it's easily sorted. Just as it has taken the realisation that we're literally eating plastic via our fish intake for us to do anything about the ocean plastic problem, it seems it's going to take something drastic happening before huge, necessary shifts are made in the satellite industry.

Beyond asking companies to de-orbit their defunct satellites which haven't burned up using space-debris mitigation technology, another approach would be to create some form of global space-traffic management system – which seems like a sensible thing to have anyway considering the thousands of bits of metal currently whizzing around in orbit.

The US Department of Defense has the best space traffic control system service in that it has clear notifications – if your satellite is due to have a close flyby with one of its US satellites, they'll alert you a few hours in advance.[18] But even with that nice system set up already, no one really has a usable global picture of what's going on. There are no predictive capabilities or notifications for all flybys from all satellites from all countries around the world. In order to have a full view of what's going on in orbit, every country needs to be involved, and all be in agreement about the best way to do it.

There are big questions we must ask about creating such a global traffic management system: Who should run (and control) it? Who pays for it? Who owns the data? How reliable is it? Who has the responsibility for building it?

Then, of course, there's the question of insurance and liability. At the moment, there's the Space Liability Convention, which is aimed more at countries rather than companies, and can be vague depending on the situation.[19] In a collision in 2009 between Russian and US satellites where both were destroyed, no claim was made. That doesn't mean that no one is going to care if their expensive satellites are destroyed through no fault of their own. And the questions remain: If a collision does happen, who should pay? Whose fault is it? If two satellites operated by two different companies based in two different countries collide, and both are following the original paths their companies programmed and planned for them, who is to blame? Is it the company that went up second, as they have, in effect, put their satellite on an orbit that interferes with an existing satellite's route – even if they didn't know there was another satellite on it as there's not a good enough tracking system in the first place? Or is it the first satellite company's fault for not making the route known to the international community? And when we've worked out whose fault it is, will insurance companies even cover the damages? (Not really – insurance companies are pretty reluctant to cover satellites, as the risk and uncertainty is so huge, due to the fact that they can't check in with a traffic management system to verify the collision!)

I was surprised to learn – after speaking to many scientists, lawyers, engineers and policy-makers in the field – that there really isn't a right answer just now to managing space debris. Or rather, there is a right answer in the form of mitigation technology and the creation of a global traffic management system, but when you understand the incentives at play in the industry that prevent adoption or even the beginning of a collaborative effort in the right direction, those answers don't quite fit.

We Are But Human

Before we go on, I want to go back to that hype-driven narrative: 'space will be different'. There is such idealism, hope and excitement attached to the idea of going into space, and building a 'human future' beyond the Earth, that we're missing the fact that the space industry has its problems just like any other. And when the problems are posed, that hype-driven narrative allows us both in and out of the industry to bat them away with a sentiment of: 'Ah we'll sort it out – it's *space* after all, the most incredible adventure in humanity's history! We couldn't possibly mess it up!' We forget that we are but humans, and the problems that an industry in space will have – and currently has – mirror the problems we already have on Earth. Some of the mechanics are different, of course, but the way governments and societies and humans work will spill over into the way the system in space runs. It's humans who, at the end of the day, are the ones controlling what gets built and done in space.

It's only by understanding how the systems on Earth work that we can start to understand how best to plan for and tackle the inevitable problems that will arise, and have already arisen, in the space sector. Space won't be different in the sense of societal systems and an in-built desire for power, and the sooner we all accept that, the sooner we can learn from our previous societal successes and errors, to best move forward. Space isn't a clean slate – we take our problems and our societal systems with us.

One of those societal systems we need to better understand, when it comes to space, is law.

Law of the Land (and Above It)

There are two key things to know about law.

First, law tends to be reactive. In 1912, the *Titanic* sank. More than 1500 people died, many simply because there were not enough lifeboats on board. In 1914, the Safety of Life at Sea (SOLAS) Convention was passed, which dictated how many lifeboats were required on ships and the number of lifeboat drills that had to be carried out, among many other safety procedures.[20] The convention has been continually updated and is still the global maritime treaty for merchant ships. The International

Ice Patrol was formed in 1914 – a body that would be tasked with monitoring and reporting North Atlantic Ocean icebergs – and still operates to this day. In essence, it took the *Titanic* sinking to force cruise ships to have enough lifeboats on board – despite the cost and the weight these safety elements added to the manufacturer's bill – to ensure everyone could be saved if the worst came to the worst.

With commercial space law, regulators want to be able to put in place rules and regulations that ensure space is both safe and sustainable, as well as keeping the economic barriers to entry low for small corporations. It's a hard balance to achieve, but the problem lies not just in finding that balance. It also lies in finding the first mover to put regulation in place, who will possibly 'pay the price' of losing out on tax and credibility if it's not done quite right, so that the second and third movers can overtake with refined ideas built on that first plan.

It's hard to form new regulation until something untoward has happened, as we need to know the steps that led to the problematic event, and then intervene at the right point with the right kind of new rules. Simply, we need to know how something goes wrong in order to then prevent it in future. Of course, people make predictions all the time about how things might play out – there's the entire insurance industry, and governments and businesses routinely have to make plans for events, such as freak floods or power outages or public relations issues, that may impact their work. But when it comes to law, there's reluctance to regulate for the unknown. With the added commercial pressure of countries and companies not wanting to have to foot the bill of a 'just in case' technology, it's easy to see why the situation is as it is.

The irony of all this is that the space industry wants laws. Companies need legal certainty that what they are doing isn't going to be made illegal later on down the line, in order to get investors on board with confidence when they are looking for money to grow. They're also saying, though, that they don't want too much specificity as they don't yet know how they'll execute certain solutions and use pieces of technology. Essentially, they don't want their hands to be tied behind their backs by laws that don't make sense in a time of rapidly changing technologies.

Soft law does exist with regards to companies in space: there are codes of conduct and guidelines and all sorts around how companies should conduct themselves in space, but they are voluntary to follow and controversial to the extent that they have stayed voluntary: there is not yet consensus on which of these soft laws should become hard.

The Outer Space Treaty (or, to give it its full name, the Treaty on Principles Governing the Activities of States in the Exploration and Use of Outer Space, including the Moon and Other Celestial Bodies) hasn't really been updated since its creation in 1967, but it's the Magna Carta of space law.[21] It has rules in it such as the prohibition of weapons of mass destruction in space, to keep the peace above our heads. The key phrase in its full name is 'Activities of States' – it was conceived with countries, not companies, in mind.

So again we're left with the issue that if one country moves first in putting in new space law directed at companies, it may be left in a position of loss. Without international agreement on new kinds of framework or treaties to get around this, the problems of the commercial space race are simply left unsolved, and quite possibly will stay that way until something untoward happens.

The second thing you need to know about law is that it relates to areas of jurisdiction.

Current legal frameworks assume things come from and operate within particular countries – which makes complete sense when you consider the fact that the planet is segmented into countries with borders.

But what about space? The area far above our heads isn't owned by anyone, so what do we do with a place that isn't part of any one country, and many countries operate within it? What does law mean in the context of satellite operations, or mistakes made by companies in orbit? Going one step further, what about murder on space hotels that have been collaboratively built, such as the International Space Station – is the alleged perpetrator to be tried under the laws of the country of the victim or the defendant, or the country that built the section of the hotel where the murder happened, or the nationality of the biggest financial backer of the hotel . . .?

To work out how we might tackle this, we can look to two places on Earth where we do in fact have a similar problem to solve – the sea and Antarctica. Both have organisations that essentially organise, regulate and control what goes on in these areas with no single 'ownership' by any country. Saying that, the International Seabed Authority, and the Antarctic Treaty System, both only have some of the countries on Earth as members, and are effective and transparent only to the extent a large international bureaucratic body can be (which, in short, has been found to be not very).[22]

There's also an organisation that is already regulating parts of space, and that's the ITU (International Telecommunication Union). It's what's called a 'UN Specialised Agency', and it coordinates radio-frequency regulations – essentially, what frequency information is beamed at so it doesn't interfere with others, and where satellites are placed. In essence, it divided up the radio-frequency slots for satellite communication to occupy, but there's been an element of 'first come first served', which has resulted in developing countries missing out on bandwidth for their internet services, due to the number of public and private companies battling over (and paying large sums for) spots in space.[23]

In mentioning these three organisations already modelled on places with no single owner, there's the idealistic and optimistic part of me thinking, 'Maybe it *will* be OK, maybe we *do* have the solutions' – but I know that this is not enough. We can't assume 'it'll be fine' when we can already see the cracks in the existing institutions that we'll likely base our space ones on. And in understanding the cracks in the existing institutions, as well as taking a deeper look into how countries really work, we can make the most of that knowledge to then build something better.

Take continent-wide or complex countrywide space agencies, such as the European Space Agency (ESA), for example. To get anything done, ESA needs to have the approval of the different member states, the national space agencies, the national governments, the European government and then the management of ESA itself. The US has three areas to contend with: state law, national law and the management of NASA. Japan, as a comparison, simply has JAXA (the Japan Aerospace

Exploration Agency) and the national government with which to wrestle.

This presents an interesting opportunity for various countries, though. India didn't inject Cold War cash like JFK did in the 1960s, but did have a strong interest in space from that time, and nowadays it has not only one of the most powerful space programmes on Earth, but also a more attractive legal framework to attract companies to work, launch or even base their headquarters there. Companies and wealthy individuals are known to flock to places such as Andorra, Barbados or the Channel Islands to avoid costly tax bills. Similarly, we could easily extrapolate out to the idea of 'space-law-positive' countries just like there are 'tax-positive' counterparts. Of course, then there are issues with tax havens in that they allow the rich to avoid putting money into their home country's tax system (and thus reducing that country's ability to take care of its citizens). And we already know about the problems arising from the tragedy of the commons, with countries acting in their own self-interest, possibly resulting in an irresponsible space industry. But understanding and confronting these realities of human society, and not just turning a blind eye to space industry problems with our internal insistence that 'space will be different', might lead to not just the shunning of bad practice, but the ability to then create better systems knowing the current one's flaws.

PLUCKY POLITICS IN LUXEMBOURG

Luxembourg is known for its beneficial regulatory system and tax loopholes, and it has – over the last few years – now turned its attention to becoming a nation fuelling the modern-day space race; and profiting from it. In 2018, it formally announced its very own national space agency, alongside a fund to help grow so-called 'NewSpace' companies.[24] Before the national space agency even existed, Luxembourg announced a law in 2017 that allowed companies and individuals in its jurisdiction to take ownership of the resources they mine in space – which echoed Obama's 2015 US Commercial Space Launch Competitiveness Act,which recognises the right of US citizens to own the space resources they obtain and encourages the commercial exploration and utilisation of resources from asteroids.[25] Luxembourg has, over the last few years, been present at many a space conference, graced the front pages of many a science magazine, and generally done a stellar job at positioning itself (and backing it up with legislation and financing possibilities that make it far more beneficial to startups than many other countries) as the nation for plucky, profit-driven space entrepreneurs.

And alongside the bid to secure outside interest, the narrative has become ingrained in those who live in Luxembourg: 'space is important for our country'. Politicians talk about space: it's part of their campaigning, in much the same way the National Health Service is revered (and used as a political tool) in Britain. Etienne Schneider, Luxembourg's deputy prime minister, leads delegations to the spaceports in Silicon Valley, and talks openly and with passion about the importance of space to Luxembourg's national interest.[26] This consistent narrative and governmental action have resulted in space entrepreneurship being part of the identity of the country, and thus it has strong domestic backing among the voting population.

Luxembourg politicians have taken the time to understand the space ecosystem, warts and all, and they are using that knowledge for national economic benefit. They have worked out that to make money and charge ahead, it's not about discussing Martian colonies or utopian space societies, but about talking to their voters about attracting the businesses currently making money to their country so that they can all benefit.

Outside of Luxembourg, politicians are less comfortable talking about space: 'Why are you spending your time talking about space law when investment in homelessness, health and education is at an all-time low?' might sound familiar. If a topic doesn't result in a vote, it's not touted as important in campaigning (or isn't mentioned at all) – and with the short-term nature of political cycles, it's the visible and immediate issues that rise to the top of the agenda, no matter how sensible (or not) that is over the long run. There's also the issue with politicians not wanting to shoulder blame if something goes wrong. If a politician talks about bringing to the fore a satellite sustainability bill in a country where the public understanding is low, gets the support and implements it, but the bill drives business out of the country in the short-term, a rival would no doubt call them out, and that politician and the party they represent would shoulder the blame. If something is not loudly discussed as part of the national narrative, few outside of the industry will know about it and support politicians making progress that is good only in the long-term. And when fewer people know, less is done – no matter how important those issues are, or how much the public might benefit from action.

When politicians don't talk about the real space industry, and when the media focuses more on Elon Musk and utopian space societies, it's clear why so few people seem to care about the issues in the satellite industry, and what might happen next – despite the fact it's arguably the most important, most influential and possibly the most problematic part of the space industry.

In many ways it's great that Luxembourg is pushing the agenda in propping up the burgeoning space industry with beneficial laws and regulation – but it's clear this is an individualistic national bid to win the race in attracting space entrepreneurs (and their income tax) into the country. We already know what happens when narratives surrounding an industry only focus on individual profit; look at the death of the rainforests due to farming expansion, or the way miners are treated in cobalt mines for the sake of our iPhone batteries.

We can't afford to avoid the harder truth in the growing space industry – states operate according to perceptions of national interest, as opposed to the interest of the global collective.[27]

Article 1 of the 1967 Outer Space Treaty reads: 'The exploration and use of outer space, including the moon and other celestial bodies, shall be carried out for the benefit and in the interests of all countries, irrespective of their degree of economic or scientific development, and shall be the province of all mankind.'[28] The fact that space superpowers such as the US, Russia, China and India have all signed this international space agreement, and yet have also developed and tested missiles that can target and successfully obliterate satellites from their rivals, suggests that, regardless of what checks and balances we have in place, individual interest and quest for power will still always come first.[29]

International relations has been studied throughout history, and researchers are turning their attention to how geopolitical theories, concepts and rules might project into the world of space power. There are many open questions – after all, we cannot predict the future – but one thing is clear: we are limited in our ability to build new theories for space as we can only project Earth-bound political, societal and commercial relationships.[30]

In short, prioritising power is all we know.

Mining Outer Space

'Mining asteroids could be worth trillions of dollars', said the headline by CNBC in 2018.[31] The first line of the article is: 'Minerals that lie in the belt of asteroids between Mars and Jupiter hold mineral wealth equivalent to about $100 billion for every individual on Earth', followed by the later admission: 'A lack of legal clarity about ownership of space resources makes things difficult, according to experts.'

Mining on Earth is a lucrative endeavour. Since prehistoric times, humans have been extracting value from the ground we walk on to provide tools, energy, weapons, building resources and a whole host of different precious materials. The industry is at the root of our construction, energy and technology industries; and the companies dominating it wield far more power than discussions in the public sphere would suggest.

Earth's resources are finite – and the fact that there are asteroids out there with huge amounts of valuable raw material such as gold, cobalt and nickel is a tantalising prospect.[32] Not only could we build a tremendous space ecosystem off the back of the wealth created – much like the mining industry did on Earth – but the value of bringing back these materials to Earth, especially as we start to run out of them terrestrially, could be huge.

There are a lot of unanswered technical questions surrounding asteroid mining, such as where the economically viable ore materials are in the space ecosystem; how easy those valuable asteroids are to get to; how cost effective it is with the cost of space missions being so high; what technology is required; what stage we are at with the technology required;

whether we know how to use that technology in zero-gravity; whether we have enough fuel to transport the tech to the asteroid, and then the tech plus the newly mined cargo off the asteroid and towards wherever the end customer is. And these questions are not necessarily problematic as such – what is science and technology if not for asking and answering questions? – but they are just scratching the surface when it comes to digging into the world of asteroid mining.

Peter Diamandis, one of the founders of asteroid-mining company Planetary Resources (now bought out due to lack of funding), said: 'Everything we hold of value on this planet, metals, minerals, real estate, energy sources, fuel – the things we fight wars over – are literally in near infinite quantities in the solar system.'[33]

Naveen Jain, founder of Moon mining company Moon Express, said: 'Once you take a mind-set of scarcity and replace it with a mind-set of abundance, amazing things can happen here on Earth.'[34]

The suggestion that space mining could be the solution to wars on Earth is – frankly – ludicrous.

The problem with those using these narratives is that they conveniently forget to include the entire history of humankind – and the fact that power and profit tend to trump global prosperity for all when opportunity for abundance presents itself.

INTRODUCING SPACE WATER

I mentioned some of the different materials that can be found on asteroids. It's easy to come to the 'trillion-dollar industry' conclusion when you start simply taking each of the asteroids and their material make up and multiplying their weight with their current price on the commodity markets. But when you do take into account the technical challenges and the cost of missions, the most cost-effective thing to mine is in fact not one of the precious metals, but instead – water.

It doesn't sound so ridiculous when you consider the cost of getting water to space. It's heavy to carry, takes up precious cargo space, and we need a lot of it to survive as humans if we are indeed also talking about putting us up in space. If we fast forward fifty to a hundred years – where missions to the

International Space Station or a space hotel or whatever else we have up in orbit are happening with the frequency of plane flights, and some vehicles and businesses and even humans are stationed in orbit (or maybe on the moon) as opposed to Earth itself – the idea that an abundance of water is available in space, as opposed to being transported up from Earth, suddenly becomes quite exciting. Not just for powering humans, but for powering the spacecraft (water can be split using electricity from solar panels into hydrogen and oxygen and can then be used as the fuel).[35]

The question is, though: who needs space water right now? Or at least, who are the customers that come first for the early asteroid mining companies? There's a sort of 'chicken and egg' situation: we don't currently have humans living in space or rockets needing refuelling in space, so there's no one there to buy the mined water and thus the space mining industry might struggle to grow. But in order to build a space industry where we do have humans living in space and we have plenty of rockets working in space and in need of refuelling, we need the mining industry to first build up its capabilities, for which it needs money.

But the opportunity in mining, at least to begin with, seems to lie in the moon as opposed to the asteroids. With an abundance of dust that could be used for concrete-like material for building lunar outposts of some description, as well as it simply being closer to the Earth than most asteroids, interest in mining the moon is growing. There are the proposals for a collaborative Moon Village emerging from the European Space Agency; there's the American Deep Space Gateway – literally that, a gateway pitstop on the moon from which to access the rest of space.[36] China is pledging to have the first moon base by 2030; India and Israel are working to get their first landers successfully on the moon; and companies such as Astrobotic and ispace are building lunar transport systems (and have customers already lining up).[37] The idea of space mining doesn't seem so crazy after all.

There is a lot of talk in articles and at conferences about how long it will take the space mining industry to get underway. There are questions around the tech, and the availability of asteroids or water on the moon available to mine, and the stability and profitability of the companies currently at the forefront. Valid questions, of course, but for me there are bigger questions.

Questions that, again, we cannot ignore by assuming what happens in space will be different from what has happened on Earth for many years – it's an industry being built by humans, and thus will have our tendencies for power-seeking built into it.

INDIANA JONES IN SPACE

Asteroids and the moon are generally thought of as 'dead rocks', worthless beyond mining them for resources. You only have to look at the way we cut down forests for land and wood without considering the CO_2-reducing effects, or the cultural significance of forest habitats (for humans, animals and all kinds of biological ecosystems) to see how humans can mindlessly prioritise the worth of resources above all else.

We don't yet know what might be the negative repercussions of mining the moon or asteroids, and while I'm not suggesting we're likely to find ET lurking, there are plenty of arguments that exploration and knowledge-gaining should come first before we exploit resources simply because we are using up the ones we have on Earth.

The question of 'where to draw the line' when it comes to powerful humans taking advantage of what's in our environment is a valid one, which is already explored in an industry built on mining: construction.

Any mention of the Crossrail project – a 73-mile high-speed railway line currently being built (it's a joke in London that the building will never end, so I'm not so concerned about that 'currently' dating this book) – is one which will emit a groan from many a Londoner. This is mainly due to the delays and closures the huge rail project has resulted in – but beyond the super-fast commuter links it will bring, the project has also played a crucial role in the world of archaeology. This happened because of a quirk of planning permission regulations that state for certain construction projects – especially the bigger ones such as Crossrail – archaeologists must first be given access to the area to quickly excavate anything they might find before the diggers destroy the ground and all the secrets within. Under the Crossrail project, for instance, they found a 55-million-year-old piece of amber, two parts of a woolly mammoth's jawbone and teeth with bacterial remains from the Great Plague.[38]

The law allows for a good balance of letting archaeologists dig without giving the construction industry a blanket ban – it preserves the needs of archaeology with the need to build. Now, I'm not suggesting that we need archaeologists to search for the remains of 'little green men' on the moon and on asteroids, but there's an argument to be made that there's plenty of science still needing to be completed on both the moon and on asteroids before we start mining for resources.

The point here is that this mode of thinking – going beyond tearing up the land for profit, and taking a moment to consider the repercussions of human expansion and prioritisation of our resource-draining behaviour – may not continue into the space industry if we do not put it into practice at the very start: now.

And the way we talk about human expansion into space suggests we're not really thinking responsibly. It's easy to spot the human desire for power when you look at the language used: the term 'colonising space' is used to refer to the idea of humans being the controlling force on places such as the Moon and Mars.

You'll see a lot of speakers, writers and academics in the space commercialisation field use the European colonisation analogy to talk about why we should go to space faster, why more investment is required and why people should be excited. You'll hear: 'We had to learn to build ships before sailing to Jamestown', 'Imagine if Columbus hadn't gone to America' and much more.

When you consider how obvious the issues of colonialism (murder, rape, racism, servitude, starvation, exploitation – the lot) are, it boggles the mind why it's remotely appropriate to use this term. Why say 'colonising' if we as a global society have all learned? (Spoiler: not all of us have.)[39]

It's not about tiny green Martians or the rights of rocks. It's about the fact that the incentives on show in the modern-day space race are mirroring those of the past. It's about the fact that power and prestige and speed are being hailed as the things to aim for, as opposed to the responsible building of an industry.[40]

There are arguments, of course, that mining space reduces the impact of mining on Earth – mining the moon could remove the need for the coal mine that is killing the Great Barrier Reef.[41] But we don't yet know what might be the negative impact of mining and expanding into space, and we're not always questioning power moves when discussing the modern-day space race. The race for power is always prioritised through-out history, and has seldom worked out for the benefit of all. We need to be mindful of that tendency when we look at what's going on today in space.

If we assume that 'space will be different' from Earth, that humans won't repeat the mistakes of the past, we will simply have been blinded by the hype. We won't ask questions about the lack of laws in place; we won't realise that profit is being put before precautions; we – if we work in the industry – will continue to build without questioning whether what we're doing is right or wrong, and who will benefit.

BUYING A PIECE OF THE MOON

When power goes unchecked – which can easily happen when hyped-up narratives prevail – we (sometimes unwittingly) create uneven wealth distribution. We're already seeing this come into play in the space industry.

Article 2 of the Outer Space Treaty states: 'Outer space, including the moon and other celestial bodies, is not subject to national appropriation by claim of sovereignty, by means of use or occupation, or by any other means'. Essentially it stipulates that no country can own anything that wasn't put in space by us.[42]

But as I mentioned before, Luxembourg and the US have instituted laws that give those under their jurisdiction licence to recreate the Gold Rush by owning what they mine. The Outer Space Treaty was written with governments in mind, which might suggest that this fifty-year treaty may not be up to the task of regulating what goes on as the modern commercial space age develops. The fact that there is regulation, though, and that nations on Earth are working hard to get more specific modern legislation in place, suggests it won't be exactly like the Wild West.

Or will it? Even if rules and regulations are set up in theory to ensure everyone gets a piece of the pie, they can also prove to be very unfair.

We've seen this in the heavily regulated pharmaceutical industry, in terms of drug prices and pharmaceutical profits; and it's no secret that the abuse and crimes that happen in the mining industry are as a result of bad incentives pushing people to prioritise money over human rights. And even if the space mining industry somehow manages to be fair in these respects, there's the simple issue of problematic wealth distribution when a 'winner takes all' power-trip mentality, cloaked in a narrative of creating a utopian society, is what is driving the companies at the forefront. Those who are most involved at the start might end up benefitting the most – so much so that they strip away the opportunity for those who come after.

Commercialising space might make the opportunity fairer. By opening it up to the free market, competition is encouraged, barriers to entry are lowered and there is simply more diversity among those getting involved. But realistically, as we've seen with so many other industries in the past, without consideration for the 'good of the many', those at the forefront prioritise their own power and wealth, and many are left behind. Making the most of the opportunity in space is really only open to those who understand the power of technology and how to utilise it. And only open to those countries where education is such that space companies can be built domestically without having to rely on services sold by those elsewhere. And only open to those with the upfront capital and security to get involved in investing their money in the first place.

Technology is often considered to widen the wealth gap – so just think what will happen when space opens up and the first movers grasp at the early wealth.[43] Those without access or unable to get involved might be left so far behind that this utopian vision of a human future in space is really only reserved for those who can afford it. One person's utopia is all too often another's dystopia.

Space isn't infinite when a winner can 'take all', resulting in the rest of us having to use a particular system or company or product, and companies and governments have monopolistic control of that access. If SpaceX proves it can create an affordable, reliable, reusable rocket, or create an effective satellite constellation that provides internet to those who don't yet have it, it will be able to take much of the market and hence reduce the opportunity for those who enter it later. In the same way a developing country feels the force of unfair creation of systems at the ITU, who is going to miss out similarly when it comes to ownership of resources in space? The race to be the first, to be the winner who 'takes all', feels problematic when 'all' refers to space.

REMOVING OUR ROSE-TINTED GLASSES

We know the issues on Earth around wealth distribution, geopolitics, lack of environmental sustainability, corporate greed and so on – we may not talk about them enough, but when pushed, we all admit that we know plenty about them and their effect on society. Can we learn from the history of Earth and stop ourselves from replicating and expanding those problems in space?

'NewSpace' is the buzzword used to describe companies and efforts in this modern era of space commercialisation – at conferences, on their websites, in the media. It refers to the fact that they are doing things differently from the big companies of the past. Yes, their business models and their technology are different, but it's idealistic to suggest they won't succumb to the same human tendencies we've seen time and time again throughout history. They must be treated as the same if we're to effectively hold them to account as they pave the way for our future in space.

When we look back at problematic times in human history, we say that we'd have said something. We'd have written about it in the newspapers, we would have protested in the streets, we'd have cared. But there's the potential for something questionable being built in the modern space industry – with inherent issues of wealth distribution, lack of access and environmental concerns – and we're letting Elon Musk's quest to be a Martian speak far more loudly than is warranted in the public realm.

The modern-day space race has the potential to democratise space in a way never seen before. There are incredible trailblazers opening up opportunities for more people, companies and countries to make the most of the resources that space does indeed afford us. Whether that's satellites connecting the world with internet all over and helping us research climate change, or, indeed, an exciting future of going to space on vacation, space is and can continue to be both hugely beneficial to society as well as the greatest adventure. Many of us entertain romantic notions about space, and understandably so.

Holding the industry to account by seeing it for what it is doesn't stifle it, and shouldn't remove our infatuation with it in one fell swoop. Through tough love, we can protect it from repeating the mistakes of the past, and better support the core mission of going beyond our planet ambitiously, fairly and for the good of the many.

The Quiet Winner in Quantum-computing Hype

....

Every March each year, I join 75,000 people from all over the world who head to Austin, Texas.[1] It's not for the rodeo or the breakfast tacos (OK, maybe a little for the breakfast tacos), but for one of the biggest technology, music and film festivals: South by Southwest. At SXSW, beyond the film screenings and music gigs, there's a conference with two weeks of talks, product launches, business announcements, government speeches, startup pitches and countless deals made – all concerning technology and its effect on the world.

There are thousands of speakers in many different venues, but the tech celebrities, the special guests and the talks on the most popular topics of that year get the coveted Keynote slots. These are the ones predominantly reported by the media, the ones that take place in aircraft hangar-like rooms with well over 3,000 seats, are broadcast to other enormous rooms nearby for those who couldn't get into the main event, and are live-streamed for everyone else around the world keen to tune in. They have had Barack Obama, they've had Melinda Gates, they've had Elon Musk, they've had Neil Gaiman – as SXSW puts it: 'Keynote speakers embody the DIY spirit, ingenuity, and entrepreneurial drive that SXSW uniquely cultivates.'

In essence, if a Keynote speaker says something at SXSW, the technology world listens.

In 2018, one of the Keynotes was delivered by William Hurley. Commonly known as 'whurley' (a username he used in the early internet days), he's a celebrity among the tech community. Having worked at IBM and Apple, and been an active open-source advocate, whurley is well-respected and regularly looked to for future predictions for where technology is going.

whurley had recently started a software company in a burgeoning new area, and was asked to speak at SXSW about the little-known world of quantum computing. The room was packed with people who had heard the term but knew little about what quantum computing is or what it could be capable of. Interest in quantum computing is increasing but it is still a topic only the select few discuss, partly because the general consensus within science is that quantum computing still has so far to go, but that doesn't stop those on the outside growing curious and hungry for information.

whurley didn't disappoint.[2] After a science primer and an overview of the current state of development, he moved on to why those in the room should start investing their time, money and attention in the world of quantum computing.

Despite whurley's disclosure that what he was talking about is still far from reality, he made big claims – even using slides with big bold phrases such as 'Fuck Climate Change' – to excite the audience. And the excitement was key, because during the presentation whurley announced his new quantum computing software startup, open for business and investment from those in the audience.

It's no surprise then that those articles written about his talk made misleading statements about quantum computing. It was not just the articles proclaiming his hyped-up points, though: many people at SXSW are there representing their companies, and when they're back at the office, these people tend to present the main learnings from the conference to their colleagues. I know this because this used to be my job.

If you were there listening to whurley's SXSW keynote, considering his main-stage credibility, you'd be forgiven for believing that quantum computing can be used to solve huge swathes of society's problems, including climate change. You'd be forgiven for thinking that the rate of the technology's progress is so fast that a quantum-computing-powered world is just around the corner. whurley, I'm sure, wasn't out to deliberately exaggerate quantum computing for his own startup's interests, but in painting an idealistic picture of the field, especially on the SXSW stage to an audience he'd have known full well would err on the side of naive optimism, he added fuel to a fire that had been heating up for some time.

It's too simplistic, and highly misleading, to say quantum computers will solve problems such as climate change. And it's, at best, conveniently disingenuous and, at worst, deliberately misleading to claim that progress is happening at breakneck speed without giving detailed context and caveats. A quantum-computing-powered world is neither realistic nor round the corner: these machines won't be replacing our smartphones or laptops, and full-functioning versions are likely to be decades away. whurley, to his credit, did try to include pieces of information which – for those in the audience already in the know – could be tied together to 'debunk' some of the big points he made. But for those who were new to the topic, a fast pace of quantum computing development and an ability to 'save the world' would have been the messages that came out loud and clear.

Scientists get frustrated when marketers, startup founders and others on the commercial side overhype their work, and they are well within their rights to be frustrated by someone else incorrectly, however accidentally, communicating the innovation. What's really troubling, though, is what happens when hype results not just in misleading speeches, but in bad commercial decisions, misguided expectations and – ultimately – the actual slowing down of development. There's a difference between simply exaggerating and negatively influencing progress.

The two most common narratives surrounding quantum computing – that it can solve any big problem either faster or simply instead of a standard computer, and that they are soon to take the world by storm – are hype. And they stick.

They stick because it's easy to misinterpret two key things about quantum computing. 'Solve any big problem' is based on the idea that a quantum computer can do many calculations simultaneously and exponentially faster than a standard computer, which isn't the full story. 'Soon to arrive' is based on the continued fast-paced coverage of the milestones companies and researchers are reaching, which is rarely put into context for those on the outside.

In short, the hype has grown, and stuck, as these narratives make sense when paired with overly simplified, incorrect ideas about what a quantum computer is and how we measure its progress.

The hype continues, though, because the field can be difficult to understand without a primer in quantum physics, which most people don't have the time to explain or listen to. And those explaining the field to the public persist in their simplified, misleading narratives, because there's money to be made in getting the masses excited, whether they understand it properly or not.

It's good to bust the hype – it makes us smarter – but there's a more pertinent problem beyond us simply not quite understanding the truth. There is a lot that we as a society might lose if we continue to get over-excited by misleading ideas about quantum computing.

A PRIME EXAMPLE

First things first: why is anyone even bothering to talk about quantum computers? In short, it's because they are advanced machines that can solve very particular kinds of hard problems, which may have huge implications for future societies. And to begin to understand the possible impact of these mysterious-sounding machines – and thus the increasing interest in them – one example we can use lies in the world of secret codes.

Secret messages are an important part of our society. There's hiding your identity so you can send emails to journalists, blowing the whistle on corporate wrongdoing, or simply buying a new toaster online without anyone stealing your credit card details in the process. It all requires encryption to ensure your information is kept firmly behind lock and key.

Encryption works very well, because at the root of it is a hard problem that no human or computer can currently solve in a reasonable amount of time. One of the most widely used methods, known as RSA, relies on the difficulty of 'prime factorisation', a problem so hard that it takes computers millions or even billions of years to get to the solution.

Quantum computers are a different kind of computer, and if we manage to create a fully working one, they would be able to solve these specific, very difficult kinds of mathematical problems. Some of these problems are ones that conventional computers are not built to solve at all, others are ones that would take them so long to solve that there's not much point even trying.

Prime factorisation – the key to cracking encryption – is one such problem which, for a fully fledged working quantum computer of the future, would be easy.

Any number can be broken down into its prime factors. A number's factors are the ones that multiply together to result in your original number. So for instance, if you wanted to find the factors of 20, you would find that they are 1, 2, 4, 5, 10, and 20 – because all of these numbers can be used in some form of multiplication to get the answer $(1 \times 20 = 20; 2 \times 10 = 20; 4 \times 5 = 20)$. Now, a prime number only divides by itself and by 1. To find the *prime factors* of 20, you have to find the combination of only primes that multiply together to get 20; 2 and 5 are the only primes in that list, and 20 can be reached using 2 and 5 like so: $2 \times 2 \times 5$. The prime factors of 20, therefore, are 2 and 5.

The prime factors of 100 are also 2 and 5, because $100 = 2 \times 2 \times 5 \times 5$; the prime factors of 63 are 3 and 7, because $63 = 3 \times 3 \times 7$, and so on. Any number can be broken down into its primes. (Although this might sound obvious or useless, this is such a deep and important mathematical fact that it is known as 'the fundamental theorem of arithmetic'.)

This is a simple process when you have numbers that easily divide by small primes such as 2, 3, 5 and 7. But if I were to give you the number 515,443, you'd soon find that it doesn't nicely divide by 3 or 7 or 11 or any of the first 126 primes. In fact, the only two numbers that divide 515,443 are themselves primes – 709 and 727 – and so you'd only be able to work this out by literally going through the list of primes from 2 up to 727 one by one and testing if they work.

So what does this have to do with encryption? In order to encrypt a message, you need to have a process hiding it so that criminals – or simply those you want to keep out – can't understand it. You also need a process for uncovering the original message for whoever you wish to read it. Using the RSA method (named after its creators Rivest, Shamir and Adelman), you have what's called a 'public key' – a number you get from multiplying two huge prime numbers together (maybe each has fifty digits); and a 'private key' – the two prime factors themselves.

You can make the public key, which is used to scramble and encrypt the information you want to hide, public but only you, or whoever you've authorised on the other end of the message, knows the private key, which is used to decrypt the secret message: 'Use this enclosed credit card information to pay Amazon for my lovely new toaster'. If anyone wanted to try and break into that message without the private key (i.e. the two prime factors), they would have to sit and factor that huge number to find them out. This would take so long – maybe trillions of years – that by the time that person managed to get hold of your credit-card information, you, and they, would most likely be dead.

The fact that prime factorisation is such a hard problem is at the root of why encryption using RSA works.

Of course, we have built computers that can do the boring, time-consuming task of finding simple prime factors for us, but when it comes to large numbers, ones that you get by multiplying two prime numbers with over fifty digits, for example, even our computers become sluggish and slow in going one by one through these primes to find the answer. There's no shortcut for this – it's literally a trial-and-error process – and because it's such a long process to solve, it works perfectly as a basis for protecting our information from people trying to steal it. In 2010, the most recent record for factoring the number with largest number of digits into its primes was set (although it is possible that secret intelligence agencies have factored a larger number without disclosing the achievement), and this was for a number with 232 digits.[3] It took the research team who cracked it two years and many hundreds of computers.[4] The public keys used nowadays are much larger than 232 digits.

Hard Problems

The fact that this problem is so hard that conventional computers cannot solve it means it's extremely useful for hiding personal information – it would be pretty disruptive to the security industry if we were to find a way to speed up the solving process. But it would be beneficial to speed up solutions to lots of other hard problems on Earth. This is why there is excitement about quantum computers: they work in a different way to standard computers and as such, might be able to solve certain difficult problems that, as a society, we need the answers to.

One such hard problem involves fertiliser. We use a huge chunk of all the world's energy making fertiliser, using a process created in the early 1900s called the Haber–Bosch process. Nitrogen from the air and hydrogen from natural gas are extracted and turned into ammonia. In order to break the bonds between the nitrogen atoms to create the new substance, we use an iron catalyst, heated up to about 450 °C and kept at an extremely high pressure. The energy required to create the ammonia – the key ingredient needed for artificial fertiliser – is therefore huge. When you think about how much food is needed by the entire world, the growth of which is powered by fertiliser, it's easy to see why so much energy is eaten up creating ammonia. For one loaf of bread, the carbon-footprint is 590 g – 40 per cent of this is from the carbon dioxide emitted not just from the fertilisation process, but throughout the chain involved in making bread – from growing the wheat to baking the dough in the oven.[5]

We know that there's a better, more energy-efficient way, as microbes in soil manage to create ammonia from the nitrogen in the air with only tiny amounts of sunlight, in a process called biological nitrogen fixation. It's the exact same equation: nitrogen plus hydrogen to create ammonia, but instead of using a chamber with high pressure and temperature to break the bonds, bacteria living in soil and plant roots 'fix' the nitrogen naturally using an enzyme called nitrogenase. We do not yet understand how they do it, all we know is that this bacteria is capable of breaking those bonds and opening the nitrogen up easily to turn into ammonia. We can't simply grab all the soil microbes and task them with creating fertiliser for us – it wouldn't create enough and they have got their own work to do – but if we could understand how they do it and copy their genius, our human-made process could be revolutionised. But to do this, to create a method that we could implement in fertiliser factories all over the world, requires simulating the chemistry of life at such a small scale, with so many different variables, that our conventional computers simply cannot handle the numbers. Nitrogenase has a substance at its heart called FeMoco, and if we could simulate that substance, we could most likely then understand the natural process. Researchers predict that the quantum computers of future could simulate FeMoco in a matter of days or months (the model assumed multiple computers were working in parallel to crunch the numbers several times to ensure accuracy).[6]

Simulating chemistry is even harder and more time-consuming than factoring a long number. There's not just one long to-do list that has to be worked through: for every bullet on your list, there's another 'sub' to-do list to be completed first. And for each entry in that 'sub' to-do list, there's a 'sub-sub' to-do list, and so on. And some of the 'sub' to-do lists link back to ones earlier on, creating interdependent loops throughout. A picture of all the things you have to do would look more like a tree that branches out almost infinitely, with random circular connections all over the place, as opposed to one long linear list. Conventional computers look at a huge problem like that, and simply cannot begin to attack it. They are built for long lists, not navigating through and choosing between infinite paths and routes. Quantum computers, on the other hand, can tackle these kind of problems head on, without having to go through a trial-and-error approach. It's not that they are faster at going through all the paths and routes, it's that their architecture and design mean they attack the problem in an entirely different way.

Discovering a new drug, and bringing it to the people who need it most, on average takes ten years and costs $2.6 billion.[7] Part of the reason it takes so long and costs so much is that the search for and comparison of molecules that work together to make a drug result in another hard problem similar to that of working out biological nitrogen fixation. At the moment, pharmaceutical companies will run billions of simple comparisons of different molecules to see which might work best as a drug on their super-fast versions of conventional computers, but they are still severely limited to using only the small molecules that these computers can handle.

If we can find a way of solving these kinds of hard problems, we may impact the world of encryption, but even bigger problems such as growing food and discovering medicine are desperately in need of better solutions. There are already many companies and research labs working on different kinds of post-quantum encryption methods for exactly this reason – it's clear they believe something disruptive is coming.[8] Something that, yes, will help humanity in so many (very particular) ways, but will also throw up lots of new challenges.

That something is the quantum computer.

OK, So What Is a Quantum Computer?

To understand what's going on in the world of quantum computing – and, thus, why hype is problematic – you need to understand exactly what it is about quantum computers that makes them special. A lot of the coverage of quantum computing in the mainstream technology press – it's still not really in the mainstream general press yet – avoids explaining the nitty-gritty of the technology for fear that the detail would turn off readers.

But keeping it too simple sometimes results in information verging on the 'incorrect', and the misleading dumbed-down descriptions give rise to poorly managed expectations about what the tech can do. By trying to explain quantum computing simply, the media tends to get it wrong.

The first main narrative in most explanations aimed at a non-expert audience is the idea that quantum computers can solve all manner of big societal problems. Beyond drug discovery and fertiliser efficiency, quantum computers are posited as some kind of technological messiah, here to clear up the problems of humankind.

The idea beneath this hyped-up narrative is: 'quantum computers do many calculations all at the same time'. And it's based on an incomplete understanding of one of the most confusing but crucial areas of science – quantum mechanics.

Quantum Mechanics – A Complete Incomplete Primer

The beauty of quantum computers isn't that they are faster than conventional computers, it's that they work in an entirely different way. As discussed, for prime factorisation, the challenge that normal computers have is that going through primes one by one takes a great deal of time. They use the 'algorithm' of trial and error. Quantum computers don't simply go through them faster, one by one, they are able to make use of a different algorithm – called Shor's algorithm – which makes use of the quantum nature of the machines. If a quantum computer can do something faster, it's not because it runs at a faster speed, it's because it can exploit a shorter route that conventional computers simply cannot. By saying that quantum computers are simply 'fast conventional computers', you're rounding them down to something a lot more ordinary than they are.

That shorter route can be explained by quantum mechanics.

One of the most famous quotes in the quantum mechanics field, commonly attributed to Niels Bohr, the Danish Nobel Prize-winning physicist, is: 'Those who are not shocked when they first come across quantum theory cannot possibly have understood it.'

It makes sense, then, that some people would argue that cutting corners in explaining quantum mechanics is the best way to bring people along on the quantum computing journey. I'd agree, but I believe it's less about cutting corners, and more about being intentional in which corners you choose to cut. If you're throwing cargo off a sinking boat to make it float for longer, it's only beneficial if you throw away the non-crucial baggage. Throw away drinking water, oars and the map to get you home, and you might as well sink then and there.

Now, quantum mechanics is the field of study that focuses on how things behave at the micro level. At the level of atoms and sub-atomic particles.

We're used to classical mechanics, with Newton's Laws of Motion, with which we can predict the arc of a thrown ball, or we can calculate how much force is needed to push a huge boulder up a hill, or know how much it's likely to hurt when we trip and fall and watch the concrete floor approach our kneecaps.

Quantum mechanics isn't quite so 'normal'. Small things behave in peculiar ways, and the key 'abnormal' idea to get your head around in the context of quantum computing is 'superposition'.

With conventional computing, the basic unit of information is a 'bit'. Inside your phones and laptops are chips made up of billions of transistors. These transistors act as on–off switches for passing information through the computer. Every command, such as 'type the letter a' or 'add x and y together', translates into a series of operations on a long string of zeros and ones, which represent each of the transistors' state – on or off. Each of these 0s or 1s is a bit, and each bit can only ever be a 0 or a 1. Never both, nothing else.

The superposition principle in quantum mechanics means that small things such as sub-atomic particles can exist as a combination of more than one state, until the point at which you measure it. If you're to take the computing information states of 1 and 0, then a bit in the quantum world – called a 'qubit' – can exist as a combination of both 0 and 1 when unmeasured. Now, this doesn't mean that the qubit is in two places at once, it means every qubit is made up of two parts – a 0 part and a 1 part. When not measured (looked at), the qubit exists as a combination of both 0 and 1; when measured it must be one or the other.

This normally translates in the general media as this: the quantum computer can have all its qubits as both 0s and 1s and therefore it can do all of the calculations a conventional computer does in sequence, all at once. Say the conventional computer has a list of actions to complete a computation – maybe it's a list of instructions on how to calculate revenue for a corner shop next year. That list of actions may amount to using four of the transistors in sequence, with their switch status as so: 0011; then 0101; then 1100. That computation might lead you to an answer of say 1110, maybe corresponding to £400,000. It's easy to then get caught up in the idea that qubits could do all these three steps in one go without the switching in between, and thus can compute and calculate answers far faster than a conventional computer – getting to the £400,000 answer in one step as opposed to three. This isn't quite right.

The reason that this is incorrect is due to the lesser-known quantum phenomenon of 'probability amplitudes'. Probability amplitudes are like the quantum mechanics' version of probability, as when you square the amplitude (multiply it by itself), you get the probability of a result. In the world we know best, probability works in such a way that if you have a probability for each option for a situation's outcome, they all add up to 100 per cent. If it's the result of a football match, there may be a 50 per cent chance team A will win, a 30 per cent chance team B will win, and a 20 per cent chance it'll be a draw. In the quantum world, amplitudes do add together, but you can have negative amplitudes as well as positive ones, so some probability amplitudes cancel each other out.

If you are to task the quantum computer to do some kind of computation, the qubits would explore all the different options for getting to an answer. At the point of measurement (the point at which you say 'all right quantum computer, give me the answer now'), you want all the amplitudes for the routes that result in a wrong answer to cancel each other out, and all the routes to the correct answer to add together and reinforce one another, as the quantum computer will feed you the 'most likely' answer first. Maybe the options for the wrong answer routes have amplitudes corresponding to 0.2, 0.3, -0.3 and -0.2 (meaning they'd add up to 0 and 'interfere destructively') and the options for the right answer route have amplitudes corresponding to 0.3, 0.4, 0.2 and 0.1 (adding up to 1 and 'interfering constructively'). In the corner-shop revenue calculation example, the qubits would have different amplitudes for each route to each different answer – maybe there are amplitudes cancelling out for routes that lead to £200,000, and amplitudes that add together for routes that reach £400,000. When you tap into the quantum computer after its calculation time, you hope that enough of the correct answer routes have been reinforced so as to allow the correct answer to surface.

This explanation is not at all to diminish quantum computers, as this amplitude quirk of the quantum world can be powerful in situations where you have so many options for outcomes it's simply impossible to check them one by one (like prime factorisation). But it's only powerful if those building the quantum-computing algorithms – the recipes the computer is to follow to answer the question you ask it – can choreograph the qubits in such a way that these amplitudes reinforce the right answer, and not the wrong one.

That is an extremely difficult thing to understand, never mind execute. The thing to take away from that, though, is that it is an oversimplification to merely view each qubit as being able to do both calculations (in both the 1 state and the 0 state) at the same time. There are certain kinds of algorithms that have certain kinds of probability amplitude choreography, which can possibly work well on a quantum computer. Which means there are only certain kinds of problems that we know of that we can solve using those algorithms on a quantum computer. It's also worth noting that quantum computers *can* do anything a conventional computer can do; whether they can do it so well is another story.

And it turns out there are only a few proven algorithms that would work well on the fully fledged working quantum computer of the future.

The first is the prime factorisation algorithm I mentioned before – Shor's algorithm. This is a quantum computing recipe introduced in 1994 by Peter Shor, which does one thing and one thing alone: factors numbers into their primes.[9] It requires millions of qubits to work.

The second most famous is Grover's algorithm, which can be used to search huge amounts of unsorted data. This algorithm, unlike Shor's, is generalisable across many kinds of unsorted data searches, but it's not exponentially faster than a conventional computer. This is again because a quantum computer doesn't change what is computable; it changes what is *efficiently* computable. For Grover's a quantum computer is faster, yes, but it's not concerned with problems conventional computers cannot do – it just solves these particular search problems more quickly. It would be like walking diagonally through a square, to get from one corner to another, as opposed to walking up one edge and then along another. Again, Grover's needs huge numbers of qubits to work, but there's a level of excitement around this algorithm because the unsorted data search is directly related to optimisation problems, which are everywhere in computer science, especially in machine learning and AI.

In 2016, a review of the field found that there were only 262 published quantum algorithms.[10] Which means there are only certain classes of problems that we currently know how to solve using a quantum computer of the future.

If we were to take a problem such as climate change, for instance, it's clear that it's not as simple as finding the representative maths problem you can feed into a quantum computer and get the right answer to pop out. Some people talk about the ability of quantum computers to model and predict the effects of global warming, based on the idea that nature follows a quantum system. It rings true to some extent (but it's not clear what exactly these more accurate models would do other than maybe giving us certainty in predictions). In other words, 'a quantum computer could just model all of nature'. It was Richard Feynman who first expressed this as the real potential of quantum computers, saying in 1982: 'Nature isn't classical dammit, and if you want to make a simulation of Nature you better make it quantum mechanical, and by golly it's a wonderful problem because it doesn't look so easy.'[11]

Others, when presenting the benefits of quantum computers with regards to climate change, talk about using them to 'solve' global warming. To do this, we first need to work out what exactly the things are within the climate-change landscape that we need to tackle. Some of those things might have something to do with simulating the chemistry of life – maybe finding a more efficient material to be used on solar panels, or maybe reducing our use of energy in the manufacturing of fertiliser, or maybe finding some kind of chemical reaction for capturing the carbon in our atmosphere. But we don't even know if all those chemical reactions exist; we know we can make fertiliser more efficiently because we know the main activating substance in nitrogenase is FeMoco – we don't know if there's a FeMoco for carbon capture. Plus, 'solving climate change' is so much more than finding new materials for renewables or carbon capture; it's also about societal demand for certain kinds of food, and the prevalence of air travel, and all manner of other human impacts. Selling a new technology with that huge promise is incredibly hyperbolic, totally diminishing the scale of the problem and – when paired with a request for investment from those not entirely in the know – irresponsible.

There's a reason that the same examples – prime factorisation and encryption, fertiliser manufacturing optimisation, drug discovery – come up when quantum computing is explained. It's because these are the problems that have some kind of quantum-computing algorithm attached to them that has already been discovered. Claiming other examples, without the corresponding algorithm to match and a disclaimer about how many qubits it would need to run, is hype.

The narrative that quantum computers can solve all kinds of big problems stops making sense when context is given to the 'simultaneous calculations' idea. It's a misinformed misalignment of the value of quantum computers, and when used in the realm of businesses marketing their products and abilities, it borders on exaggeration or even lying.

THE QUBIT RACE

The second main narrative that tends to get hyped up, based on an incomplete understanding of the weird and wonderful field that is quantum mechanics, is that 'they are soon to take the world by storm'. This narrative tends to proliferate because if you follow the mainstream coverage of the field, you'll read about IBM, Google and Microsoft all competing to build a quantum computer with the largest amount of qubits. The 'race' to the 100-qubit quantum computer is often touted as similar to the modern-day space race, and by giving those on the outside an easy-to-understand industry measure – 'number of qubits' – the coverage becomes much simpler to follow and get excited about.

The problem is that number of qubits isn't actually a great way of measuring progress of the quantum-computing industry. This comes back to another key quantum mechanics quirk called 'entanglement'.

Entanglement is a property that can be seen in the world of the very small, where two or more particles interact in such a way that each one's quantum state cannot be described without considering the other's. In other words, if you measure one qubit, the whole system is measured alongside, and thus all the qubits collapse out of their superposition state and into their classical state. If you want to know if one of your qubits is either a 0 or a 1 after a period of computation, when you look at it, all the qubits are looked at too, causing them to be either a 1 or a 0 (as opposed to a combination of them) too.

This makes sense, and is very powerful, when a quantum computer has many qubits all entangled together, working in tandem on a problem. Because each qubit can represent many different states and can work together with many other qubits that also have this property, the number of solution options explored goes up exponentially with every qubit added to the system. It's what makes a system of qubits so special and so useful in solving these specific forms of problems.

What we must also realise is that measuring the system – and in the process of doing so, collapsing the whole thing out of its quantum state – doesn't always happen intentionally. In fact, it's extremely difficult to keep the system entangled with all the qubits in superposition for any great length of time. It also gets harder the more qubits you add.

This collapse of the system happens as a result of 'decoherence'; when the quantum system comes into contact with the material world. This could be in the form of sound, vibrations, light, radiation or heat – things that are extremely difficult to control living on planet Earth. To try to control these outside influences causing quantum computing systems to collapse, some designs of quantum computers need cold temperatures, others need extreme vacuums, while some others are surprisingly robust to the external world.

This decoherence problem – also referred to as 'noise' – gets bigger the more qubits you add. Just like the more qubits giving us more computational power, the more qubits give us more noise with which to contend. Some researchers, though they are in the minority, believe that the quest to eliminate noise defies fundamental theorems of computation, and that we'll never get to a point where full-scale 'universal' quantum computers without noise are a reality.[12]

This is where error correction comes in, and the quest to build a quantum computer with less noise. Correcting that noise is vital if we are to trust the answers the quantum computers give us. The challenge here is that error correction works qubit by qubit, so the more qubits in the system, the more error correction you need to overcome.[13]

To try to get around the problems with error, the key is to build what's called 'logical' qubits. The idea is that you put together 1000 qubits that are as error-corrected as possible and in conditions to keep noise at bay to form 'one' trustworthy qubit, which can be relied on for calculations. Each individual physical qubit inside this group is called a 'stabilising' qubit, and when we see the numbers being touted by companies such as IBM and Google, this is what we're counting. For a fully fledged working quantum computer of the future which could run Shor's algorithm, we need a quantum computer with thousands of *logical* qubits. But each logical qubit needs a thousand physical qubits to stabilise it, meaning we need a quantum computer with millions of qubits in total.[14] We haven't yet proven, beyond theory, that we can even create these logical (error-free) qubits, meaning the quantum computers all the big players are working with right now are highly prone to mistakes.

Suddenly the idea of counting qubits, and seeing the progression from IBM 'raising the bar' to 50 qubits, to Google 'reclaiming the quantum computer crown' at 72 and so on, feels a little less like exponential growth.[15] That's not to diminish these scientific and engineering feats – which they truly are, this stuff is incredibly hard to do – but when positioned like a race for the public to tune into, there's a risk those on the outside might miss the fact that the field is still in its infancy.

The number-of-qubits measure doesn't work because it doesn't capture the variance of quality within it. It doesn't tell us whether the qubits referred to are logical, stabilising or even low-quality types that shouldn't really be called qubits in the first place.

In order to shortcut understanding of the full complicated picture, a new form of measure needs to take the place of the current misleading default. Getting to grips with all the science and all the players throughout the whole industry is too big a burden to put on everyone who wants to have an opinion or make a decision on quantum computing, so a new, better shortcut is required.

Welcome to the NISQ Era

By the time this book gets into your hands, we'll still be in what's called the NISQ era of quantum computing. NISQ stands for Noisy Intermediate-Scale Quantum, and it describes the stage quantum computing research is currently at, and will maybe stay at for the next decade or two.

In January 2018, John Preskill of Caltech wrote the paper titled 'Quantum Computing in the NISQ Era and Beyond', which brought the term out of obscurity and into the popular discourse in the field.[16] It outlined the difficulties in getting to a future full-scale noise- and error-free quantum computer, but at the same time, acknowledged the sudden recent increase in interest in the quantum-computing field, especially from big businesses and startups. The paper marked a call to action of sorts for those in the field: to take the progress made so far, and find applications. It wasn't defeatist, it was almost a call for creativity beyond building an idealised final version so that we're not waiting decades to start reaping the benefits from the incredible advances already made. It gave the field focus, alignment and navigation.

Preskill called for NISQ-era algorithms, hardware and approaches that focus on the 100-qubit quantum computer, giving the field a new way of talking about advances, and segmenting the research and commercialisation efforts into near-future and far-future efforts. Simply, this new concept made it easier to work out which timeline people were focusing their innovations on.

Not only is this a shift in the way people in the field talk about their current goals, but it's a way for those on the outside to better understand what's going on, by asking a simple question: 'Are you working on something for the NISQ era or not?' If non-experts can evaluate the stage at which companies and researchers are at, different questions can be posed to those behind the innovation. For instance, if a company isn't working within the NISQ era – in other words, it is working on solutions based on the need for a full-scale quantum computer of the future – questions can then be asked about how they intend to fund themselves for the next decade, or even beyond, until that comes to be, or what problems exactly they are aiming their technology at. If a company is working within the NISQ era, then the questions can be more targeted towards who the current or soon-to-be customers are, what problems they can currently work on, and how they are proving their claims right now.

Part of the reason hype propagates in this field is because it's hard to know the full picture, and a misleading measure – number of qubits

– has been used as a shortcut to understanding, which has then led to incorrect views on the state of the industry.

Having a framework for evaluating the state of an industry helps cut through hype, even for those who aren't experts in the field. Due to the nascent nature of the physical qubit abilities as well as the yet unproven engineering feat that comes from creating logical qubits, simply counting qubits doesn't quite allow for a full understanding of the field. If those on the outside – the politicians making budget decisions, the investors making funding decisions – get overly excited about the qubit count, lesser technologies are the ones more likely to be backed.

The best remedy to hype is knowing what the hype is made of. Finding shortcuts is key to understanding the whole, fast, as long as it is not a shortcut to the wrong destination.

And that's exactly what happened only a few years ago.

THE D-WAVE HYPE

The need for understanding what constitutes *quality* quantum computing research and development, and what happens when misinformation comes into play, can be encapsulated in the story of one company infamous in the industry for – shall we say – conveniently under-explaining.

In February 2007, Dr Geordie Rose took to the stage at the Computer History Museum in California and presented to the world his quantum computing company, D-Wave, and its first 16-qubit proof-of-concept processor.[17] By 2009, the company had raised $65 million from investors including Goldman Sachs.[18] In 2011, it released its 128-qubit D-Wave One, described as the 'world's first commercially available quantum computer'.[19] In 2012, it raised another $30 million, this time from investors including the CIA and Amazon founder Jeff Bezos.[20] In 2013, D-Wave landed a contract with Google and NASA, and sold one of its devices for $10 million to Lockheed Martin.[21] In short, D-Wave rode the wave of hype and excitement that grew around its perceived speedy growth and commercial positioning in an otherwise academic-focused field.

Questions had been asked in the academic community, right from the first announcement in 2007, but the media hype seemed to peak in 2013 with so much perceived success displayed by the company in terms of technical progress and investor confidence. Articles in the *New York Times*, Engadget and the IEEE blog (Institute of Electrical and Electronics Engineers) spoke of D-Wave's supposed success in trials and presented D-Wave as the one-to-watch in bringing quantum computing to the market, fast.[22]

The biggest question being asked among experts, though, was about D-Wave's qubits. It seemed impossible that it could be coupling so many qubits and keeping them entangled in the quantum state for enough time to perform computation. It turns out D-Wave wasn't working on a universal quantum computer, but instead a so-called 'quantum annealer'. Because of this particular technology focus, this meant that its computers could only work on certain types of problems: optimisation problems. These problems are worth working on in the commercial world; if we can speed up solving optimisation problems, we can essentially search through mountains of data faster, meaning companies that work with huge sets of information could theoretically be more efficient. The justification in investing huge sums of money in a quantum annealer system to work on these kinds of problems would not only be to solve the optimisation problems, which conventional computers tend to be sluggish at or not capable of, but to be able to get a formidable speed up, allowing those in need of an answer to employ them quickly.

Academics got to work running simulations to work out how much faster the D-Wave computers were than conventional computers. A 2014 study showed that D-Wave's machines not only weren't capable of providing that significant increase in speed, but also weren't even capable of solving the problems any faster than existing conventional supercomputers already in the market.[23]

It was at this point, around 2013, that the voices asking the questions and calling out D-Wave as over-claiming and failing to correct misleading articles began to get louder.[24] Saying that, in 2015, Google announced that its D-Wave machine outperformed a standard computer by 108 million times, which made positive headlines again.[25] What it failed to make clear was that the machine outperformed in the particular 'problem' of being able to simulate itself, a task specifically designed to be easy for a D-Wave machine and hard for anything else.[26] Having a quantum annealer capable of beating a standard computer at the task of simulating its own functionality, as opposed to doing anything of real-world use, arguably shouldn't have made headlines. In fact, the same results showed the computer working on another problem, one which is more useful in the real world, and they showed that a standard computer was faster. But the announcement, with its lack of context and tantalising clickbait, shot D-Wave back into the realm of 'worthy investment' hype.

In 2018, D-Wave was still talking about its ability to 'maybe help on issues like climate change', without the broader context of its work always being presented alongside those kind of claims.[27] The company say it doesn't matter that it isn't working on solving Shor's algorithm and building a universal quantum computer, with its CEO saying, 'To me [it] isn't an important application space, not something we particularly want to be involved in – you know, invalidating all of modern cryptography.'[28] In 2019, D-Wave hit headlines again with its 5000-qubit quantum computing platform announcement – again relying on that big number for attention, as opposed to proof of error correction, noise reduction or actually speeding up optimisation problem solving.[29]

By this point, it was common knowledge within the quantum computing industry that D-Wave's technology had to be referred to as quantum annealers due to how misleading simply referring to their qubit count was, but mainstream articles covering the space still missed out this fundamental caveat, prompting many on the outside accidentally to believe we are fast approaching a full-scale working quantum computer (as opposed to the truth: that we are years away, and D-Wave isn't even working towards that anyway). And in 'overview of the industry' articles and presentations, D-Wave is still sometimes used as an example of the exponential growth of the industry, without all the historical and scientific context.

Saying all this, D-Wave *is* one of the few companies in the quantum computing field that is actually shipping product – people are buying its machines in full, or buying time on them to perform calculations remotely; NASA, Lockheed Martin and Volkswagen are all still customers, having spent between $10 million and $15 million with the company; Ford signed up too in 2018.[30] The opportunity to speed up optimisation problems is indeed there – especially considering the amount of data we as a society are now able to collect. There's talk of using the D-Wave machines for optimising traffic, for matching patterns in vast genomic datasets, and for analysing risk in the financial markets. The quantum annealers, in theory, may have some applications in pattern recognition and data optimisation in the years to come, but until they are categorically proven to be better than conventional computers, the narratives around D-Wave are still hype.

All of this is to say that widespread use of quantum computing, never mind a universal quantum computer, is not here yet. And when metrics are used out of context to showcase the progress of a field, the wrong conclusions can easily take hold. The hype around the field would suggest we're close to being able to do all manner of incredible things with these machines, which misses the point that we're in a limited era of development, and misleads those on the outside into thinking that investment and interest now will lead to huge gains in the near future. This simply doesn't seem to be the case by any stretch of the imagination. Considering the ramifications of a universal quantum computer on the world of encryption and cyber security, the idea that a quantum computer is just around the corner leads to headlines suggesting their development presents an 'imminent threat' and other such fear-driven narratives.[31]

My underestimation of the near-future gains in the field may be incorrect. I can't predict the future, after all. We *might* see some crucial piece of research surface in the next few years that makes huge progress on the noise or error problems, or that presents a NISQ algorithm capable of having an impact on the world at scale. But overestimating quantum computing's rate of development based on shaky ground isn't just idealistic and optimistic, it's a misinterpretation of the current state of knowledge.

THE QUANTUM-COMPUTING PLACEBO

The D-Wave story is a telling example of where the venture capital world and the academic world collide – they have completely different standards of what's allowed to be said. Academics have to be extremely careful with their claims; so much so that the message can often be lost in the caveats. Companies tend not to be held to account in the same way, particularly at the start of their journey when vision and ambition are lauded – nay, required – by those funding them.

There's an argument to be made that without D-Wave's early hype, the company wouldn't have been able to build to the point it is now – where it *might* offer some form of value. If it had been upfront about what exactly its annealers did and their potential, it might not have been able to get the investment to develop into something worthwhile. But on the other side, there's an argument to be made that by being untruthful, money which could have been invested elsewhere was 'lost' to D-Wave, and other – perhaps better – developments might have suffered.

It's kind of a self-fulfilling prophecy. D-Wave announces that it has built the world's first commercially available quantum computer, it gets investment and, using that money, it goes on and builds something different, but arguably still useful. If the goal is to create more useful things, then surely the 'white lie' can be forgiven? If you give someone a placebo drug and they feel better as a result, does it matter that you've conned them into thinking whatever their ailment was has been treated by a real drug?

Hype can be thought of in much the same way: it's needed to bring support to people with big, complicated dreams and ideas. If the final result of hype ends up having a positive impact, the hype doesn't tend to be remembered as problematic; that only happens when things go wrong. To some, hype in quantum computing is positive, as interest and excitement drives investment, which – even if it's going to a company under misinformed circumstances – still bodes well for the industry as a whole.

To others, though, hype in this industry could be devastating, as there could be other problems at play when you promise the world to those on a schedule.

WINTER IS COMING

Startups working on quantum computing have been popping up all over the world, with most forming in the last three years. This is great news in terms of speeding up progress as it pushes the older, wealthier players such as Google, IBM and Microsoft to continue investing huge amounts into their own efforts, it allows for more people to be working on more problems – both on hardware and software – across the field, and it means more of the research happening in university labs is being spun out into the market.

With the growth in sheer number of startups and the hype surrounding the field giving the impression quantum computers can solve all manner of problems and are just around the corner, many investors are keen to throw money at these startups. There are those investors who do the due diligence and are spotting brilliant scientists and companies, and understandably want to get a piece of the promising action (and are happy to wait a long time for their returns). But investors don't always invest with that same objective in mind. Some investors want to be seen to be investing in the latest technologies, so they can attract the largest sums of money from their clients to their funds. Some corporate innovation teams need to show that they are 'future-proofing' their companies by investing in the science and technology of tomorrow, and spending $4 million on a startup investment or in creating an innovation team is a drop in the ocean of large-scale corporate marketing budgets.

With any new advancement, there will always be those early companies who – consciously, or not – gain money, fame and growth simply for being early to the market. Some of these companies do become key players in future markets, but some will be pushed too fast by the eager investors to whom they have sold a dream in order to get their companies off the ground in the first place.

Some of these startups are exposed as having overhyped to get investment, some fail for normal and understandable reasons. The problem with the former is not just in losing their investors' money; it can change how the non-expert players see the field. There's a risk that over-eager startups jumping on the quantum computing hype bandwagon, and making big claims to investors not doing key due diligence, brand the field away from being something exciting and 'in fashion' to something overhyped and full of con men.

This shift has happened, and is still happening to some extent, in the AI field.

In 1984, the term 'AI Winter' was coined. Roger Schank and Marvin Minsky – two reputable leading AI researchers who had been part of the growth of the AI research field in the 1970s – warned businesses that enthusiasm surrounding the technology had got out of control. They argued that there had been too much hype, followed by disappointment and criticism from investors and governments not getting the results they were promised, and that there were soon to be funding cuts, which would then slow future progress.

Sure enough, three years later, the billion-dollar AI industry collapsed and serious research struggled to get the same levels of attention and investment.[32] It wasn't until twenty years later that renewed interest for the technology moved it forward once more. That renewed interest is currently very high, to the extent that the term 'Artificial Intelligence' is regularly thrown into a startup pitch to investors without actual implementation – in fact, it was found in 2019 that 40 per cent of AI startups in Europe don't actually use AI at all in their products.[33] Again, the hype is driving investment without the due diligence necessarily being conducted. Being seen to be investing in AI, as opposed to supporting real efforts, still carries prestige and commercial marketing weight.

There's a concern that a hype cycle in quantum computing may soon come. Investments in the startup space have blossomed in the last few years, with private investors joining the government grant funding pots, and there is still huge appetite for progress and commercialisation of the science.[34] If we're really to benefit from this astounding science, we cannot afford for interest to peak too early and for funding to dry up before the larger real-world problems begin to be properly worked on.

THAWING THE ICE

Some argue that there's nothing to fear with investment winters in quantum computing, as the field is so tight-knit and most of the money and focus concern serious applications, which makes sense when you consider the stage that the technology is at. They'll say that the shorter, overly simplified articles and the presentations at Davos and SXSW do little to affect the positive investment in the field. In other words, they say that the field doesn't need the interest of the masses. They'll say hype, to an extent, was and is required within certain circles in politics and finance in order to keep the ball rolling, but getting the general public on board is simply an exercise in science communication, not in broader industry influence.

Take the Internet of Things (IoT) – the industry concerning all manner of hardware connected to the internet, and thus, other hardware. Internet-connected fridges that tell you when you're out of milk; Fitbits that track your heart rate and send it to an app on your phone; wi-fi enabled kettles for . . . well, who knows what? Around 2014, IoT was pretty trendy. There were articles all over the technology-industry press and the exciting futuristic ideas started leaking out into more mainstream publications. Companies were talking about their 'IoT strategy' and many a marketing team was working out how they could incorporate some kind of internet-connected device into their next TV ad. It had become a big enough conversation topic that people could start to make jokes about it (there's a great Twitter account @internetofshit).

But time passes, people become less interested, and trends shift. Nowadays, most of the narrative around IoT in the consumer-tech world is poked fun at; a company coming out with a brand new IoT product isn't really seen as innovative, but rather, out of fashion. IoT – in the eyes of the tech-savvy general public – isn't that cool any more.

Inside the industry, however, things couldn't be better for IoT. Factories with machines connected digitally are hugely increasing yield as they are able to make efficiencies and run automatically like never before. Business buildings such as offices, manufacturing plants and data centres are saving money on energy, as they have installed sensors and connected up their air-conditioning, lighting and heating units to dynamically shift throughout the day, preserving cash and carbon at pace.

If you look at IoT in the industrial world of factories – as opposed to tech-gadget consumer products – the level of general interest doesn't seem to correlate to the volume of real-world implementation and adoption. Industrial companies are buying; it doesn't matter if it's not 'cool' or in the top-ten tech lists of the year.

The same argument could be made about quantum computing – it can be spoken about both in the business and startup communities (with their tendency to overhype for big investment valuations and shareholder satisfaction), as well as the general public interest sphere, without damaging the field's continued growth.

Should We Keep Schtum?

But with AI, those conflicting narratives *did* have a sizeable impact. The funders who put money on the line based on hyped-up promises grew impatient, and started to withdraw follow-on funding and interest, so much so that the rate of growth of the industry slowed down.

It's worth pondering, then, whether we should 'take the risk', as it were, in talking about quantum computing as part of general discourse at all. If overhyping the technology might lead to poor investment choices and impatience at a subsequent lack of results, which possibly runs the risk of an 'investment winter', and the hype might not even be needed to keep the technology on track within the industry anyway, then why bother taking the risk? Are we putting the development of the industry at risk by trying to simplify and thus give out bad information, and thereby shooting ourselves in the foot?

You could argue that there are two ideal options when it comes to communicating complex ideas and then making a decision off the back of them.

The first: give everyone perfect information. That means making sure absolutely everyone has the ability to fully understand the message, in all of its complexity, and that those giving out the message do it thoroughly, at scale and consistently. Of course, this would not only mean revamping the education system, but require a lot of time and money to effectively get the message out to all.

The second option goes as follows: assign a few decision-makers whose job it is to have all the information – perfect information – and come to a conclusion about next steps. No one else is told anything, and everyone else has full trust in those making the decision that it will be the right one. Of course, this would mean everyone agreeing on what a 'right decision' means – a question of personal morals – and everyone would need to truly trust those in charge of deciding the course.

I'm not under any illusion that either of these is a viable option, but if they are at opposite ends of the spectrum, what does our reality – which sits somewhere in between – look like? Should we be closer to perfect information for all, in which case we get more people understanding the full picture and thus a greater chance of not-bad decisions being made? (I say 'not-bad' as 'good' implies we all agree on the best course of action, and that's a personal decision for each and every one of us – all technology has trade-offs, as we saw in the batteries chapter.) Or should we be closer to assigning expertise and trusting those designated to make good on our faith?

The Quest to Understand the Universe

There's a quiet winner among the hype in quantum computing. While the financiers, corporates and startups push each other to keep researching and investing more and create a demand that drives the industry faster, the researchers focusing on fundamental science rejoice.

Science is financed predominantly through government-funded grants – pots of money set aside for scientists to apply for via research councils. There's pressure on the research councils to fund science that leads to national economic wealth and job creation – the public funds these pots through taxes and so there's a pressure to spend the money in such a way that it benefits society. That all sounds reasonable, except when you consider some of the most impactful discoveries came not from a solution-based approach, but rather from curious experimentation.

The World Wide Web was originally built as a tool for automated information-sharing between the particle physicists at the experimental laboratory in Geneva, CERN, and the rest of the collaborating scientists around the globe. Electricity, CRISPR, X-rays and countless other examples of breakthrough science happened because of curious people wanting to understand the world; asking big questions about the nature of life, the universe and everything. If the innovators at the helm of those breakthroughs had to fill in a grant application form in order to get funding, they would have had to stipulate how it would further society – and at the time there was no way of foreseeing the significant impact they would have.

Nowadays, though, science funders are regularly blamed for creating a culture of 'incremental' research – research that simply builds a tiny bit on top of what was previously done, so that results can be generated quickly, and reports on effectiveness of funding can be written. There's a severe lack of funding going into fundamental research. A former chief economist at the International Monetary Fund goes as far as saying that the drop in government-funded research over the last fifty years is one of the key drivers behind the decline in US productivity growth.[35] There's also been a decline in corporate-funded research labs, such as the famous AT&T Bell Labs, which worked on lasers, transistors and all manner of programming languages. The hesitancy in funding such curiosity-driven research by both governments and private funders is because there's no telling if the research may eventually result in something useful beyond the new knowledge itself. The funders don't have sufficient trust that they'll get bang for their buck, as it were.

But in quantum computers, and their eventual ability – in the years to come – to simulate life at the quantum scale, fundamental researchers have found a tool for discovery. A full-scale universal quantum computer – with its error-corrected algorithms and noise-cancelling engineering – should, in theory, be able to simulate the behaviour of the molecules too big for our conventional computers. Understanding the behaviour of these molecules and particles and other super-small matter means understanding how life itself 'works'. It was this purpose Feynman was originally referring to when he spoke about quantum computers in 1981.[36]

The biggest opportunity in quantum computing is merely being able to get to grips with life at its most basic – like we've never been able to before – and maybe, just maybe, use that ability to better understand the universe we live in.

You'd think that would be reason enough to win investment, but knowledge doesn't always equal profit, and so the overhyping in the commercial fields of quantum computing, in some sense, is crucial for the development of basic science.

An 'investment winter', or another form of slowdown, wreaks havoc in academia, so there's cause for thinking that the hype is not so bad, and that those benefitting from misinformation in the short term might be good for broader society in the long term. It gives those, such as myself, who seethe at the idea of great technology and science being taken advantage of by those overhyping in the name of profit, a form of solace that those businesses are unwittingly helping those in it for bigger, more fundamental reasons. I don't like that those people get rich, but I really don't like the idea that the potential of quantum computing could be lost due to a simple lack of interest.

Balancing Hype with Knowledge

As has become evident in this chapter, in the world of quantum computing, there are two hyped-up ideas that can be found throughout the media coverage, startup pitch decks and corporate thought-leadership presentations: we're told that quantum computers can solve all manner of big societal problems, and that a fully working universal quantum computer is just around the corner. These narratives not only misrepresent the industry and leave those on the outside with a poor understanding of what's going on behind the scenes, these ideas are also allowing for sub-standard efforts to be funded, and covered by media – sometimes at the cost of those working on more robust, long-term solutions. Exposing this hype, by understanding exactly why it is hype, and why those saying it say it (because it serves as a shortcut in explaining a complex topic, and directly benefits plucky entrepreneurs with the nerve to overclaim), means the industry is safer still from the risk of an 'investment winter'. Hype and misinformation in the field can be rationalised in that the hype may be needed simply to get those with money and power on board, with fewer questions asked. The hype and misinformation in the commercial world may also be a good thing in keeping up the momentum with regard to the progress of fundamental scientific knowledge.

All of this is to say that hype in quantum computing – and in all areas of science and technology – is complex. It's not universally good or bad. But if we want to avoid the negative consequences of misinformation, as well as more mindfully promote the areas of science with arguably greater long-term societal benefit (as opposed to letting it accidentally happen behind the scenes), an awareness of what hype is and how it happens is crucial.

We the people aren't so slow or stupid or unable to think critically that we need to be somehow tricked into getting on board with technology, but if we continue to be fooled by hyped-up narratives, those doing the communicating will continue to treat us as such. It's worth bearing in mind that, on the whole, those dumbing down quantum computing are doing it with the best intentions – those working in the field are passionate about this technology and desperately want it to exist in the world, even if it means low-key fooling of those listening to them.

In future, though – with another technology, or for another business, or from another government – this overhyping may not be done with such pure intentions in mind.

PART THREE

Nearing

Need Versus Want in Brain–Computer Interfaces

....

There are only two devices that have appeared in science fiction that, from the moment I first saw them, I desperately wished were real.

The first is the 'Point of View Gun' from the film version of Douglas Adams's *The Hitchhiker's Guide to the Galaxy*. It's a gun which, when used on someone, causes them to see things from the point of view of the person who fired it at them. We're told that the gun was commissioned by the 'Intergalactic Consortium of Angry Housewives', who were tired of ending every argument with their husbands with the phrase: 'You just don't get it, do you?'

The second is the headjack of the synthetically grown humans in *The Matrix*, a USB port to the characters' brains if you will, where any knowledge or skill can be loaded up into someone's mind instantly, just like a program can be installed on a computer.

Of course, I'm not alone in wanting a technology that can inject more knowledge into my head or get across what I'm thinking and feeling more effectively. You could argue a technology that enhances communication, knowledge and understanding is what's missing in our current state of political affairs. But until recently, such a technology was never considered worthy of mainstream focus.

In April 2017, Tim Urban, writer of the popular blog *Wait But Why*, posted a 30,000-word article titled 'Neuralink and the Brain's Magical Future'. It told the story of Elon Musk's latest venture: a brain–computer interface (BCI) startup called Neuralink, which, as Urban put it, was creating 'a wizard hat for the brain'.[1] This device was to be inserted under our skulls, tap into the signals our neurons emit, decode them (and thus 'read our thoughts', as it were) and communicate them both to external technology devices and other people. It was a 'wizard hat' for controlling the world around us, for enhancing our neurological abilities, for short-cutting and avoiding the perils of talking and writing to one another, for keeping humans relevant in a world increasingly dominated by AI ... and it was to be in the hands of regular people in only eight to ten years.

As with any Elon Musk-related news, the message spread throughout the technology industry like wildfire. Headlines at the time read: 'Elon Musk Wants to Connect Brains to Computers' (*Guardian*); 'Elon Musk's Neuralink Wants to Boost the Brain to Keep Up With AI' (TechCrunch); and 'Elon Musk to Plant Computers in Brains to Prevent AI Robot Uprising' (*Independent*).[2]

In the same month, at Facebook's developer conference F8, ex-head of the US Defense Research Projects Agency (DARPA) Regina Dugan announced that Facebook was investing in brain–computer interface technologies to allow its users to input text directly through thought. And with the increased interest that Musk and Zuckerberg brought to the space that year, the third Silicon Valley vehicle in the BCI race made it to mainstream media: Kernel, founded in 2016 by Bryan Johnson (also founder of payment system startup Braintree, which was sold to eBay in 2013 for $800 million), would be focusing on brain training and memory enhancing.

The year 2017 was when brain–computer interfaces went from science fiction to science fact, at least in the popular discourse. It was the year that our interaction with BCIs shifted from occasionally seeing those 'miraculous' videos where a paralysed person controls a robotic arm with wires protruding from their head, to the soon-to-be top of our Christmas lists gadget. It was the year that the final slide in futurist pres-entations on the future of tech switched from 'the future is voice' to 'the future is thought'. It was the year a technology, which has not yet fulfilled anything like its huge potential in the medical field, shifted to a trendy consumer product of the near future.

It was the year brain–computer interfaces shifted from something certain people desperately need to something the masses desperately want.

A History of Melding Mind with Machine

The idea of brain–computer interfaces wasn't new in 2017. In fact, the phrase originated in a 1973 research paper by the University of California's Jacques Vidal.[3] Vidal proposed that electroencephalography (EEG), where electrodes placed on the scalp track brain-wave patterns, could be translated into commands for external technology. It was then that the idea that people could control devices such as artificial limbs and robotics with their thoughts started to be taken seriously.

The 1990s, though, was when the first devices started to emerge that proved our ability to merge human with machine, particularly for those who have lost some form of cognitive function, when researcher Philip Kennedy implanted the first BCI device into a woman paralysed with a severe case of motor neurone disease. (In 2014, to the astonishment of the BCI community, Kennedy paid to have electrodes implanted into his own brain so he could experiment on a healthy person – himself. He was driven by frustration around the lack of pace in the industry and funding available for his speech-decoding research.)[4]

As the device technology improved and more initial animal testing was completed, understanding increased around what sort of implantables were likely to work for patients in the future. It was soon clear that in order to get BCI devices out of the laboratory and into the healthcare system, even just into larger-scale clinical trials, private money would have to flow into the system through commercialised versions of the technology. As such, in 2002, a Brown University spin-out called Cyberkinetics was formed to start gathering both the financial resources and regulatory permissions for a first-generation neural interface device, called the BrainGate system.[5] In 2004, the FDA awarded Cyberkinetics an Investigative Device Exemption (IDE), meaning it could start trials in hospitals with paralysed patients.

By 2007, though, the money started to dry up at Cyberkinetics, and the company was left with no choice but to divert its attention away from BCIs to other medical devices, and in 2008, it had spun off its device manufacturing to another company called Blackrock Microsystems. The BrainGate Research Team was then formed – a consortium of university research teams and teaching hospitals which, in 2009, received from the FDA an IDE for the updated BrainGate2 system to continue furthering their research, funded by federal and philanthropic sources.

The BrainGate system consists of a sensor that is implanted into the brain attached to an external decoder, which is then connected to some form of prosthetic or whichever technology the brain is to send signal commands to. The sensor is a tiny array with one hundred hair-thin electrodes arranged in a square of rows and columns, which is implanted onto the surface of the brain to read the electrical impulses of the neurons inside. The patient essentially has a hole cut out of the top of their skull, with a box sitting atop the entry point and a long wire taking the signals from the array to be decoded and translated for the external technology. The system isn't perfect, but it's the most widely used invasive system that has made it into clinical trials.

In 2013, Barack Obama announced the BRAIN Initiative, a $4.5 billion pot to be spread across twelve years of research into brain–computer interfaces, with 2016 to 2020 being all about technology development and validation, and 2020 to 2025 focused on applying those technologies to finding out more about the brain. Many different projects are funded under this scheme in the US, from brain-mapping projects to new interface technologies.[6] Also in 2013, the European Union launched its ten-year programme, the 'Human Brain Project', to build the research base further in Europe across the whole of brain-related science, with a total estimated budget of just over €1 billion.[7]

Nowadays, the Silicon Valley startups are a part of the broader BCI field alongside the researchers, so a shift towards finding commercial applications and practical applications for the technology has begun, and the race to bring the brain–computer interface to market is now in full swing.

MAKING SENSE OF THE FIELD

There are many different forms of BCI, which differ in invasiveness and accuracy, and focus on different outcomes for the individual patients (as they historically have almost always been patients, as opposed to healthy individuals). There's the least invasive method, EEG, where electrodes are placed on the scalp, all the way to various forms of implants that are embedded into the top layer of the brain itself.

Most of the efforts focus on so-called neuroprosthetics, which aim to restore damaged hearing, sight or movement by connecting the brain signals to an external device. The most widespread BCI device is the cochlear implant, which restores hearing, and much research is focused on getting paralysed patients to move external devices, such as robotic arms, with their thoughts. There is also work being done not only to decode the brain signals to command the external device, but for such devices to have sensors on them, which can then return sensation to the person by sending signals back to the brain.[8] Maybe a brain-connected prosthetic foot could communicate the softness of a new carpet, or how freezing cold the bathroom tiles are in the morning, for example.

There is also research in restoring the ability to speak or express thoughts, mainly achieved by having the patient type text in some way. For example, the researchers may have the patient imagine moving a cursor around a keyboard displayed on a screen next to them, navigating to each individual letter to spell out their message. The motor cortex is an area of the brain better mapped than most, meaning that the signals corresponding to 'move left' or 'click mouse with finger' are somewhat easier to decode accurately, hence this approach.

The other main established medical use for brain–computer interfaces is called deep brain stimulation, where large electrodes are implanted directly into the brain (through a hole in the top of the skull), which thrust electrical impulses into the brain. It's crude, almost the equivalent of bashing the TV to get it to work, but it can have remarkable results in reducing seizures for patients with epilepsy, reducing tremors for patients with Parkinson's, and reducing symptoms for patients with obsessive-compulsive disorder.

THE SHIFT FROM NEED TO WANT

The BCI field has predominantly been focused on research and relatively rare medical cases. Nowadays, though, you could split the industry into roughly five areas: exploratory research, medical use, digital interactions, experimental wellbeing and merging our cognitive power with that of our computers. The first two focus on those who need BCI technology; the rest are making the most of the wave of consumer demand.

From a research perspective, there are many different questions about the brain that need to be answered. Which bits do what? Which signals mean what? How can we measure signals effectively? How can we avoid opening up people's skulls? The list goes on. DARPA is one of the key funders of BCI research, so even though much of the testing happens on patients with some form of neurological disorder (as they are the ones prepared to be experimented on), the questions being funded might also have some ultimate form of applicability to military methods or rehabilitation of veterans, mainly: enhancement of cognition and prosthetics.

Focusing on patients isn't just about finding willing guinea pigs, though: the social and commercial opportunity in treating or curing neurological conditions with brain–computer interfaces is huge. There are many medical reasons for wanting to read what's happening in the brain and connect it to technology. From deciphering the thoughts of a patient with locked-in syndrome so they can communicate externally, or calming the painful thoughts for those with clinical depression, to controlling bionic limbs for those who are paralysed so they can walk, or getting rid of severe tremors so those people can simply drink a glass of water or write their name, there are millions of patients around the world in need of restored or augmented movement or senses.

As we're seeing nowadays, though, it's not just patients who are keen for BCIs to hit the mainstream market. Most of us rely on our phones and our laptops daily for communication, for work and for leisure, and with that rise in digital reliance (some might say addiction), interest in better ways of interacting with our devices is increasing. The idea of just being able to think as an input, as opposed to typing or using our voices, is exciting for many a technophile. For gamers and those interested in technologies such as virtual and augmented reality, brain–computer interfaces might offer a better way of playing or working in these digital worlds, using gestures and thoughts to best manipulate the virtual space.

There's a sizeable wellness crowd wanting to meditate more effectively, a consumer demand for 'brain training' and other such productivity-enhancing activities, and an increased interest in 'biohacking' to push our bodies and minds ever further. All this makes for an almost impatient experimental wellness group eagerly following the BCI headlines. As such, crude consumer 'brain-reading' devices, as they are marketed, have already hit the market, as well as unregulated 'brain-enhancing' devices, seemingly enhancing memory by essentially zapping the brain with electrical current. This is the area where the gimmicks and the pseudoscience roam free.

The final area where BCIs are increasingly touted as revolutionary is in merging human beings with computer systems. When you consider the reams of data from all over the world that our phones and computers search for, sort and analyse, and consider for a moment merging that power with the natural powerful cognition we have tucked away inside our skulls, the idea is that those powerful systems could come together and make us more powerful beings, capable of making sense of the world like never before. It's this idea that technophiles, particularly those who follow Musk's view of the world, find tantalising.

But with the rise in conversation and excitement about BCIs as a result of the recent Silicon Valley interest, the emergence of consumer products hitting the market, as well as a disconnect between what is being spoken about in academic circles and in the public sphere, there seems to be a skewed understanding of what BCIs can really do.

WHAT CAN WE ACTUALLY DO WITH BCIS?

The main player in the academic brain–computer interface world is DARPA, the US military's research and development outfit. With the sizeable funding it has pumped into the area since the 1970s, it has been pushing the agenda for decades with cash, in both academic and industrial research settings. If you've ever seen one of those 'miraculous' videos shared on social media, where a paralysed patient is controlling a robot arm with many wires attached to their head in some form, it's most likely DARPA-funded researchers behind it.

There are many questions still to be answered in the BCI field, which makes for a very active research community focusing on various different elements, all bringing us closer to that accurate, least-invasive, useful communication link with the brain.

How to Read a Brain

There are various different technologies that can be used for BCIs. To choose which is best, there are generally three characteristics to consider. First, there's scale: how many neurons can be recorded simultaneously? Second, resolution, of which there is 'spatial resolution' – how close to individual neurons does the technology get? – and 'temporal resolution' – how close to 'real time' can the measurement record? Finally, invasiveness: researchers can decide whether they want something that is more invasive (in other words, inside the skull or inside the brain) and therefore likely to have a higher spatial resolution; or less invasive but easier to test across more individuals, with possibly more scale (think whole-brain cap versus small implant) and more satisfying as the ultimate end-product.

Within the field, there's the popular analogy of a football stadium. If you stand outside the stadium, you'll hear the crowd 'go wild' when goals are scored or when other exciting points in the match occur. This is equivalent to listening to and measuring neuron signals using the non-invasive electroencephalography technology. EEG is the most common technology used for BCIs; you might have seen people in those ridiculous-looking swimming-cap-like contraptions, with plugs all over, wires sprouting from each one. Sometimes the plugs are attached directly to the skin all over the head. Sometimes a load of gel is mixed into the hair underneath the cap to make the connection stronger. Either way, the electrodes are placed on the skin and listen for neuron signals. EEG lets you know if *something* is happening, but it doesn't really tell you *what*. Its spatial resolution, therefore, is low.

If you are standing inside the stadium underneath one of the stands, maybe next to the food stall or the bar, you'd be able to work out which stands are louder than others, maybe letting you know which team has scored if you know which sides of the stadium are dedicated to home and away fans. This is equivalent to the more invasive ECoG (electrocorticography), which, like EEG, uses electrodes across the surface to take readings, but instead of putting them on the surface of the skin, they are put on the surface of the brain, beneath the skull. It has both better spatial resolution than EEG and better temporal resolution, as it doesn't have to contend with the skull getting in the way of the signals. It's invasive, though, as a hole in the head needs to be cut, but as it sits on top of the brain, or is slid under the top coating called the dura, it doesn't really go inside the brain and is thus considered only slightly invasive.

If you want the equivalent of being on the steps of the football stadium stands, so you're able to work out which rows were louder than others, and maybe be able to tell where there are rowdier groups of fans versus more subdued families, you'd have to go into the brain with implants. Instead of discs that sit all across the surface, scientists use micro electrodes: hair-thin needles that are inserted into the brain matter. With this approach, called 'local field potential', you can only measure a small area, meaning the scale is low, but the spatial resolution is much higher than that of ECoG and EEG. The array that the BrainGate Research Team use is simply one hundred of these hair-thin electrodes.

The technology that effectively lets you stand next to an individual and hear them yell their particular chant is called 'single-unit recording', and this is when the point of the microelectrode is made sharp like a needle so that it can pick up the signal from a single neuron. Of course, this has no scale whatsoever, but has the highest spatial and temporal resolution, with the same invasiveness as local field potential methods.

Where the football stadium analogy falls down is when you realise it's not a big enough metaphor. When you consider that a standard football stadium holds about 50,000 people, whereas the brain has 100 billion neurons, the difficulty in being able to listen to each one of those neurons simultaneously, and thus receive and decode 100 billion clean signals without any noise from neighbouring neurons, becomes clear.

Scientists, therefore, are working on finding the BCI that somehow manages to combine high scale, high resolution and low invasiveness in order to create the most accurate and hence most useful technology.

There are the researchers designing 'neural dust', basically sprinkling tiny sensors all over the brain; and a team is working on a silk-based system, where the silk would be embedded with the tiny sensors and then not only be layered all over the brain, but would slip into all the folds and crevices of the brain's matter to get even better coverage.[9] There are the researchers working in a field called 'optogenetics', who are trying to find ways to genetically modify neurons so that they are responsive to light and, as such, so that their signals can be read and controlled literally with the click of a switch.[10] And there are the teams working on a so-called 'neural mesh', lined with tiny electrodes, which could be injected directly into the brain with a syringe.[11]

GETTING AROUND THE KNOWLEDGE GAP

What we do know how to do is record highly coordinated activity from the movement of millions or billions of neurons but, as you can imagine, this is very noisy, and only certain kinds of activities can be discerned from so many neurons firing at once. It's not just about finding ways to unearth the signal of every neuron, though. There are many things researchers can do without knowing exactly how every part of the brain works and how each neuron is firing.

For example, we can pick up the signal related to the movement of the left side of the body, but not the signal of individual fingers moving. We can see if the eyes are open, but not exactly what image is being seen. In research scenarios, you can look at different psychological states (for example, stress levels) as the way the brain is connected at a high level changes depending on these states, and thus affects the signals of millions or billions of neurons at once. This mass-listening approach is how we can look at sleep stages, as each stage affects the whole brain's activity. If we want to decode more specific signals with confidence, we have to record over long periods and repeat the process many times.

BCI devices aren't yet sophisticated enough to 'read' thoughts per se, and so the methods focus on patterns of change in the brain's activity, which researchers, and decoding algorithms, try to match to a particular thought. The researchers essentially have to find clever ways of getting around our scant knowledge of the inner workings of every part of the brain.

Decoding a message as a whole, such as, 'My name is Gemma', is far harder to do as we don't yet know what the signal looks like for 'think and then say this particular phrase', and so typing out the letters one by one by using our understanding of the motor cortex has traditionally been the method of choice. Remember the 'move cursor left' example, where the patient patiently directs a cursor around a keyboard? It works, but it's very slow. About seven-words-per-minute slow. That's slower than what can be achieved using other technologies such as eye-tracking, and many people who are paralysed can still move their eyes, meaning the invasiveness of a brain implant that can only achieve this speed isn't really worth the bother.[12]

There are some interesting experiments in changing the design of the keyboard to make it easier and faster to navigate, for instance, making it circular or changing the QWERTY set-up so that the letters in common words are closer or incorporating a better predictive text, but, ultimately, typing out thoughts isn't the same as saying them out loud.

In 2019, neurosurgeon Edward Chang at the University of California went one step further and recorded the area of the motor cortex that deals with commanding the vocal cords, another movement-based signal.[13] His team were then able to run those signals through a computer model of the human vocal system to synthesise speech digitally from those brain signals. The results of the study showed that the digital output was more intelligible when based on signals from someone speaking out loud as opposed to mouthing the words, meaning until it is tested on patients who cannot speak who will be asked to imagine speaking, we cannot know if the imagined signals will be easy enough to decode. Still, it's a clever way of thinking about how to get thoughts out of the brain and communicated in the real world, and it illustrates how complicated deciphering the right signals can be. Every movement, idea or action we take isn't just one neuron firing, and the research challenge isn't just simply matching up each neuron to each action. Saying something to someone involves initial idea sparking, deciding to say it out loud, choosing the tone in which to say it, breathing regulation, movement of vocal cords, movement of eyes or hands or whatever other body language you use to get the message across, and so on. Finding and isolating the right bits of the brain to read, and maybe one day stimulate, is a very difficult task.

WHAT SCIENTISTS STILL NEED TO SOLVE

There are a lot of technical challenges when it comes to brain–computer interfaces, which researchers are working hard to try to overcome.

For example, if we can get to the point where we can read and understand the signals of the brain, and we want them to feed into some external machinery, maybe a bionic arm, we need to be able to process the signals in close to real time (so that there's not an awkward lag between you thinking 'shake hand' and the arm moving accordingly). To do that processing, we can either send the data to an external computer to compute and decode, and wait for the signal to come back, hoping there won't be much of a delay, or we can add size and power to the implant itself. The problem with the former being the transfer rate lag; the problem with the latter being that this requires more electronics on board the implant, which means there'll be more heat produced. We can only increase body temperature by about 0.5 or 1 °C, so we're then limited by how much extra we can add.

Then there are the different kinds of electrodes that can be used. Maybe there are ones that may allow for a better connection or a lighter material or something else that makes the BCI work better. But technology transfer of new medical materials from the science lab to the market is expensive as the burden of proof for new materials to be used in the body is so high. This, of course, makes sense: we don't want to be sticking poorly tested materials into our brains willy-nilly. But it means that regulation is slow and pricey, and so we have to stick with old electrodes that were approved years ago, while we wait for the new ones to come to market.

There's also the fact that if we're using the implanted array, we don't even use all one hundred of the electrodes, but only about ten at a time, as the rest are there to account for noise. There are questions around how we can pack them better or come up with more efficient electrodes, or use AI to better analyse the signal output and gain efficiency in that way.

Also, everyone thinks differently. The way our brains compute and command might differ from person to person, meaning that a BCI often needs to be used by an individual for quite some time before it learns what is correct and incorrect signal decoding for them specifically. My signal for 'pick up and drink that water' might not be the same as yours.

The almost daft-sounding but completely obvious problem is that putting metal things into the most complex and misunderstood organ in our bodies is quite hard. Wires going through the skin, as well as a big connector which often sits on top of the brain, can easily cause infection. Putting a rock-solid electrode or an array of them into or on top of jelly-like brain matter means it wobbles, so the signals change day to day, there's mechanical mismatch, and those small movements of only microns of distance mean that a technician might need to boot it up and calibrate it every time you use it. And the body might simply reject the implant.

MAKING BRAIN–COMPUTER INTERFACES PAY

Perhaps the biggest challenge in the world of brain–computer interfaces, though, is how to get them to market.

There are many steps to getting medical devices to patients. There's the initial experimentation, then the many clinical trials, then awaiting regulatory approval, then finding the right distributor or pharmaceutical company to sell it to the right hospitals or doctors in each country, then making sure insurance companies cover it or national health services agree to reimburse the costs. At the time of writing, there are only two legitimate medical BCIs being used by patients in the clinic and at home at scale: cochlear implants for deaf patients and deep brain stimulators (DBSs) for Parkinson's patients. Everything else is either in the research lab or is not a proven medical device and thus is more of a gimmicky consumer product.

Cochlear implants and DBSs are arguably quite simple technologies, they have been around for some time, and they have a large enough medical population that makes them worthwhile to keep offering to patients and dedicate time to improving. The biggest chunk of the costs has already been paid in getting them approved, so now the companies that sell the devices and treatments can keep innovating and marginally improving the technology without breaking the bank.

The same cannot be said for the majority of the technologies described in this chapter. They are almost all at the early experimental stages, even the technologies from the early 2000s, because the cost of taking them from the lab to the market is so huge. It means the researchers will keep trying new and different approaches to these devices and technologies but won't actually be able to then focus on creating a particular product (such as robotic arms for patients with a very particular condition), which could eventually be sold to patients.

It's not for want or trying, though; it's simply extremely expensive and finding someone to foot the bill isn't easy.

First of all, most of the BCIs are calibrated specifically to each individual, so creating one product that can scale to many people, and thus make the research and development costs pay off over time, is no mean feat. Advances in machine learning can help here, by training the devices faster and finding more patterns across everyone who has used them to improve the technology as a whole, but that doesn't eliminate costs, it merely reduces them.

There's a big difference between the huge research grants that are awarded for early-stage research and the private money that is then required to take innovations into the real world. For starters, the incentives for the funders are completely different.

Government puts money into basic research on the condition that results are published and maybe, at a push, publicised. When the scientists apply for money, they have to show that they are increasing the knowledge base of the field, but they also need to do something realistic in a short timeframe so that they can record results quickly for the funder. There are many that argue that this system is far too short-term in its thinking, and encourages only incremental research as opposed to big-thinking science that potentially changes the course of progress in a field. There's also a case to be made that funding science like this, without the need for real-world application at the end of it all, targets the wrong kind of innovation and doesn't incentivise anyone to work hard at creating something that ultimately can make it to market.

For venture capitalists, angel investors or corporates looking to acquire teams or technologies, they wish to put their money into something that can reinforce that investment a few times over in maybe three to seven years' time. That means that if they hand over $10 million to get a piece of research through research, development, clinical trials and regulatory processes, they'll realistically only put money into something that looks to either have a huge customer market at the end of it, or is priced so highly that a smaller customer base can repay their investment, provided insurance companies or national health services agree. The technology is hard to scale, the customer base is fragmented (in that there are many unique cases as opposed to lots of cases of the exact same ailment), so the payout at the end in the eyes of the investors doesn't match up with what is needed as an initial investment to get the technology to market.

There's another side to the private-sector funding problem too, where products that have not been proven to work, but look like they will sell, get funded. If investors believe they will get a return on investment, they will invest.

And indeed there's a whole segment of the BCI market where this has already happened.

The Rise of the Consumer 'Brain Readers'

Taking control of your health using the power of technology is an enchanting idea. The reducing costs of personal data-tracking devices, the rise of the direct-to-consumer DNA- and blood-testing startups, the abundance of information at the end of a Google search, and the general societal trend towards personalising and optimising your routine, your diet, your exercise and your self-care specific to your wants and needs, all make for a hungry consumer base just waiting to be sold the next big thing making bombastic personal wellbeing promises.

Enter: brain wearables. Or brain-training-stress-relieving-sleep-induc-ing-sport-optimising-weight-losing-meditation-hacking sleekly designed expensive headsets with little or no science to back up their claims.

'Brain-stimulating headset aims to replace sleeping pills', the *Telegraph* wrote about an EEG headset that monitors brain-wave signals and plays soothing music.[14] The company's website, where it sells its €399 device, reads: 'Less time falling asleep, less waking up in the night, more time sleeping. Dreem 2 is the new way to get the sleep you need.' A fantastic copywriting job, considering the modern-day focus on us all needing more sleep; this is not just for insomniacs, this is for everyone who is tired on their morning commute.

Also tapping into the zeitgeist, another consumer EEG headset company goes with 'Technology Enhanced Meditation' as its tagline to sell its mind-only (£199) or mind-and-body (£239) EEG device to those swathes of people being told they should meditate but find it tough (in other words, those in the developed world who have ever consumed any form of modern self-help content in books, magazines, podcasts, videos or Instagram posts).[15] 'The world's best meditation tech just got even better,' said Mashable.[16]

The problem with these consumer EEG headsets is that they take the BCI technology with the lowest resolution, and make it even worse. With hair in the way, far fewer electrodes on board (the EEG headsets in the lab have about fifty, most consumer headsets have around four) and no conductive gel being applied to improve the connection, the readings are very rough. The companies manufacturing and selling these devices can get away with this as they tend to focus on the brain-wave measurements that happen across the whole brain, such as sleep states, but there's no guarantee that these crude devices aren't picking up the electric current produced by muscles in the body, or the voltage generated by heartbeats, or the electric signals from your nearby phone. Also the insinuation of much of the coverage is that these headsets edit or stimulate your brain somehow, when in fact they read signals and show you them on your phone display. They only stimulate your brain in the same way that reading this book does. The idea is that by *seeing* your brain waves in real time, you can edit your behaviour in response, but without the science to back up these claims, with respect to the particular device being used, in the particular setting they're being used in, and the health situation of the person using it, we don't really know if this is true.

FROM LISTENING TO ZAPPING

Alongside the EEG headsets are the transcranial direct current stimulation (tDCS) headsets, which are also gaining increased attention and big cheques from the wellness, health and brain-hacking crowd. Instead of reading brain waves, tDCS involves passing a weak electric current through the brain and is rooted in the recent science of neuroplasticity – how the brain's connections adapt to change in experience and signals. tDCS is meant to enhance that process, meaning that the synapses that are associated with the new thing we learn fire more readily, and so the idea is that by using tDCS, you can learn more quickly. The method has been tested in labs for conditions such as depression and anxiety and has shown some promising results, but the method can also cause skin burns and even bursts of rage, and the effects of overuse aren't well understood.[17] Some studies even show that tDCS reduces focus, and some scientists are proposing that the devices might simply be powerful placebos.[18]

Despite all this, there are now many consumer tDCS devices on the market, for anyone to buy online at the click of the button and the input of credit-card details.

There's Halo Sport, which claims it can help athletes boost their training, and that it 'works by applying a small electric current to the area of the brain that controls movement, putting it into a state of hyperlearning'.[19] It has been promoted by passionate amateur athletes in outlets such as *Men's Health*, and has even been adopted by professional teams such as the Golden State Warriors.[20] The second generation headset costs a tidy $399. On its website, the company refers to many papers on tDCS, which have no relation to its device, and to two independent, peer-reviewed papers that show positive results for the Halo Sport, but one paper tested the device on only nine men (four or five of whom received a placebo) and the other on only ten men (five of whom received a placebo).[21]

Telling people they can improve their sporting edge is one thing; telling people they can lose weight is another. Not strictly tDCS, Modius targets an electric current at the vestibular nerve, just behind the ear to, as the website claims: 'positively influence the hypothalamus and brain stem to reduce cravings, decrease appetite and make you feel fuller, quicker'.[22] The company launched the device after a crowdfunding campaign on Indiegogo in 2017, and it went on sale for $499. The *Mail Online* ran the headline: 'Could this headset help you lose weight? Cutting-edge device sends signals to the brain to shed body fat WITHOUT changes in diet or exercise.'[23] The research papers cited on the Modius website only refer to general research about the vestibular nerve, with no specific proof related to its consumer device.[24]

FALSE, OR MISLEADING, ADVERTISING

With EEG, there are no real negative health repercussions from using the device, other than maybe stress and anxiety as a result of a frustrating lack of progress, but the danger with tDCS and similar brain-stimulating technology is that it is on the market without FDA approval and thus without confirmation that it is safe to use.

Judy Illes, a neurologist and neuroethics professor at the University of British Columbia, published a study in 2019 that examined the claims made by companies selling brain wearables directly to consumers.[25] Her team found that of the 41 direct-to-consumer device websites they examined (twenty-two were consumer EEG headsets and nineteen delivered tDCS), only eight included a link to a relevant, peer-reviewed research paper and twenty-nine had links to general research that wasn't specific to their product. The others relied on user testimonials or in-house, unverified research. Nine of the companies made health claims specific to conditions such as PTSD, clinical depression, chronic pain, motor neurone disease and ADHD; but only one device cited FDA approval.

Illes and her colleagues pointed to examples such as 'The drug free way to learn to focus', 'A smart headset helps you to achieve peak mental fitness', 'Versus provides a path to wellness through brain exercises', as well as the bordering-on-illegal: 'Our non-invasive, bioelectronic platform effectively targets autonomic nerve pathways important in a number of disease processes while providing superior safety relative to pharmaceutical interventions.'

With the rise of mindfulness and self-care in the general public discourse, the idea of being able to 'hack' (or shortcut) your way into optimising your brain using technology (as opposed to using the traditional method of practice), BCIs for personal wellbeing are desirable, no matter how proven the methodologies and technologies are. The allure of the 'quick fix', of the 'technology will save us' mindset, is tantalising.

As Illes put it in the paper's conclusion: 'Continued vigilance to the claims landscape for brain technologies is especially important as this market captures the imagination of a neuro-obsessed world.'

People taking control of their own bodies and their own health has brilliant repercussions in taking burdens off health systems if prevention is leading to less need for treatment down the line, but it can also encourage questionable choices and poor decisions with respect to so-called 'alternative medicines'. Everyone should have the right to choose what they want to do with their own bodies and health, but misleading claims around wellness products and approaches don't give individuals more choice if the information is biased. Promoting these brain wearables with over-claims, misleading scientific 'evidence' and the logos of the media outlets they've been mentioned in, without making clear the limitations of the technology and the many caveats in the scientific literature, takes advantage of the lack of time, lack of expertise and lack of confidence people have to verify the details.

This is not choice, this is misleading hype for the benefit of profit. And with the market for direct-to-consumer neurotechnology products expected to exceed $3 billion in sales by 2020, it seems this is an industry fuelled by the simple desire and wealth of the consumer as opposed to the efficacy, safety or validation of the product.[26]

Just because we want something and are willing to pay for it doesn't mean it's a fair, trustworthy and transparent market, or that the device should even exist in the first place.

There's a difference between creating something to fulfil the needs of the market just to make money, and seeing the demand of the market and putting effort into creating the real thing people actually want. And while researchers focus on those who need BCI technologies for managing debilitating conditions, there's another set of people focusing on big science but for those consumers hungry for a self-help device.

SILICON VALLEY ON THE BRAIN

As mentioned, in 2017 Elon Musk released news of his latest venture via the popular blog *Wait But Why*.[27] With a $27 million input either from himself, trusting investors or a combination, he built a team and a mission for Neuralink, a BCI startup focusing on human enhancement and communication to keep us relevant in the age of AI. To ensure we stay economically viable, Musk claimed that we need a 'merger of biological intelligence and machine intelligence'.

Musk's vision is to be able to tap into the brain directly to communicate thoughts to the outside world, to our devices and directly into the brains of other people, as he believes that the speed at which we convey information digitally, whether through typing or voice technologies, is 'ridiculously slow'.[28] The answer, in Musk's view, is to create embeddable chips that allow for 'consensual telepathy' between human beings, as well as bidirectional communication with external technology allowing for quick translation of thoughts out, and many experiences, senses and brain-optimising stimulations back in.

Neuralink hasn't released much information on how exactly it is going to do this, other than to suggest it is looking to create a neural lace, similar to the neural mesh created by a team led by Harvard professor Charles Lieber.[29] Musk has hinted at an approach that would involve injecting the neural lace into a person's arteries to then unfold inside the vascular system to interact with neurons.[30] That might be less invasive in terms of cutting holes in heads, but inserting a mesh of sensors into the capillaries inside the brain isn't exactly non-invasive either, never mind proven to be safe.

And Musk isn't the only one throwing huge sums at the BCI field. As briefly discussed, in 2016 Bryan Johnson of Braintree fame put $100 million into his own startup, Kernel.[31] Initially focused on an implantable memory chip, Kernel pivoted to a less invasive approach in the hope of bringing a product to market faster, for the benefit of healthier individuals who presumably won't want their skulls opened up. Johnson says that they are now more focused on clinical use, but this is likely to be only the first stage in order to generate revenue, before pursuing more of the *Matrix*-like brain training and *Minority Report*-like device interaction that he initially touted as the ultimate goal. Kernel's team, however, haven't yet disclosed the specifics of the technology or what makes their approach so different to what's happening in the government-funded labs.

The importance of wealthy entrepreneurs such as Musk and Johnson entering the space, though, can't be better illustrated than by the founders of the company that Musk bought the Neuralink trademark from. Electrical engineer Pedram Mohseni, a professor at Case Western Reserve University in Ohio, and University Distinguished Professor Randolph Nudo, of the University of Kansas Medical Center, started NeuraLink (with a capital L) in 2011 and trademarked the name in 2015.[32] As *MIT Technology Review* reported, they were working on an electronic brain chip that might ultimately help those with brain injuries, and when they were contacted by a stranger offering tens of thousands of dollars for the company name, they accepted, none the wiser that it was Musk behind the sale. Despite the pair's research providing early data that backed up the concept, at least in theory, raising investment from private investors was proving difficult.[33] There was no guarantee that their device would be able to serve a big enough market to make the initial investment pay off, and no guarantee that it would even work. Early-stage science can be difficult to evaluate in terms of commercial application and so raising money in the so-called 'deep tech' space can be hard without a level of risk-taking attitude on the part of the investors.

People such Musk and Johnson don't have the same problem, when they can put large sums of their own money into the early stages of research and development or, using their profile and past success, convince investors to give them money. If, after that money dries up, they have progressed far enough so that they can prove they have product-market fit or that their devices are likely to work, raising bigger rounds of investment should be a lot easier. They are more likely to traverse the so-called 'valley of death' that many startups fail to overcome when they can't get that initial injection of high-risk capital from trusting investors.

FACEBOOK'S BRAIN-TYPING MISSION

The Silicon Valley entrepreneurs aren't just starting new companies, though: one company is on a mission to embed BCI into its existing product, and that's Facebook.

Facebook is focusing its efforts specifically on getting thoughts directly from the brain into the Facebook platform. As Facebook's Head of Research & Development, Regina Dugan, said at their announcement in 2016: 'What if you could type directly from your brain?'[34] She explained that non-invasive scanners would do 'optical imaging' and translate that neural activity into text, and that it would be 'a silent speech interface, one with all the speed and flexibility of voice, but with the privacy of typed text . . . Better yet, with the ability to text a friend without taking out your phone, or to send a quick email without missing the party.'[35] It was unclear exactly to what technology Dugan was referring, stumping neuroscience experts in their reactions, particularly when she said that the goal is to get to a hundred words per minute typed from the brain. (Bear in mind that the technology of the time could, at best, do seven.) But with a sixty-strong team of scientists and technologists already in place, it seemed Facebook really was going all in.[36]

Facebook's CEO Mark Zuckerberg added to the rationale for the 'need for speed', saying: 'The problem is that the best way we have to get information out into the world – speech – can only transmit about the same amount of data as a 1980s modem. We're working on a system that will let you type straight from your brain, about five times faster than you can type on your phone today.'[37]

There was some emphasis on the technology being developed to help users with disabilities, but the bulk of the mission seemed to be linked to simply allowing people to type their thoughts faster into Facebook. This doesn't seem like a use-case that would prompt many people to sign up for an implant, but with the non-invasive technologies of the time not being able to read general thoughts and instead focusing on the motor cortex, the promise of Facebook and Dugan was vague to say the least. Some experts pointed out that even if Facebook finds a way to improve the words-per-minute, the amount of mental energy it requires to concentrate on directing a BCI to type a hundred words, possibly letter by letter, would be exhausting.[38]

In 2019, Zuckerberg spoke at Harvard about the role that technology plays in society, in an interview with Harvard Law School professor Jonathan Zittrain.[39] When Zittrain brought up the constitutional right to remain silent, in light of a technology that allows eavesdropping on thoughts, Zuckerberg simply said: 'Presumably, this would be something that someone would choose to use as a product.' You would think after everything Zuckerberg had been through since the 2016 elections, the Cambridge Analytica scandal and all manner of revelations about the issues of privacy and consent on his platform, he wouldn't reduce a fundamental question to such a flippant answer.

Indeed, at the end of Dugan's speech in 2017 she said: 'Is it a little terrifying? Of course . . . If we fail, it's gonna suck.'[40]

To me, this seems like a chillingly nonchalant consideration of the downsides the technology might bring, from a company well known for not considering the societal consequences of its innovations.

BLINDED BY FAME

Elon Musk has an interesting take on the responsibility of Neuralink: he sees it as protecting us; a technology we need as we move forward into an ever-more technology-fuelled world. In an interview with Axios he said: 'The long-term aspiration with Neuralink would be to achieve a symbiosis with artificial intelligence and to achieve a sort of democratisation of intelligence, such that it is not monopolistically held in a purely digital form by governments and large corporations.'[41] By talking about BCI technology as some kind of human-empowering device that we all will need in order to stay relevant, the tone shifts from a fun sci-fi consumer product to some kind of tool or requirement of being a twenty-second-century human.

For Facebook and Kernel, where the message is about convenience and human augmentation, it's hard to see their ideas as anything more than the 'cool tech gadgets' of the future. And the cost benefit of having companies meddle with and gain access to our brains starts to feel out of whack. The hype around having these unbelievable devices in the hands of tech-lovers who can afford them can sometimes drown out the reality of what these devices are and the motives of those who are creating them.

There's no doubt that criticism of famous tech icons such as Zuckerberg, Musk and even the late Steve Jobs has become much louder in the public sphere. But there are still many people, a lot of whom work in the technology field and are thus actively building the future we're all to live in, who look at these entrepreneurs with fanatical awe. The 'cult of the entrepreneur' phenomenon means many people consider so-called 'unicorn startup' founders as mythical heroes. 'Yes, they may have flaws but, come on, they're just trying to make the world a better place. What are *you* doing?'

These 'spiritual' leaders are fiercely defended by their followers, and have a level of immunity from criticism from those within the industry who fear the wrath of the fans, despite many of them being proven to be wrong, or even dangerous.[42]

These are tech products, which ultimately could have huge societal impact, being sold as a dream device by those who are idolised and fiercely followed, without transparent details of method or regular media scrutiny. We must see past both our urge to unquestioningly idolise as well as our strong desire to have something we want, but might not need.

Is Silicon Valley Good for The Field?

On one hand, you could argue that it doesn't matter that keen techno-optimists are idolising those who are building incredible products. Surely it's better to idolise someone who is out there building the world somehow, instead of a controversial YouTube personality or dishonest Instagram influencer? Where the problems arise, though, is when those fans are blinded by their appreciation, and encourage technology development that distracts from the bigger issues in this field, i.e. patients who need the tech to have better lives.

The excitement and interest Musk, Zuckerberg and Johnson bring to the field, though, could be an overall positive for those working on basic research or indeed life-changing medical applications. There doesn't seem to be any evidence that the hype the Silicon Valley entrepreneurs bring to brain–computer interfaces is negatively impacting everyone else: it doesn't look like the medical field is bearing the cost of the hype, idealism or bad actions of these new companies, at least not yet. Similarly, you could argue that Professor Stephen Hawking had a much bigger impact on public consciousness of brain–machine interfaces, but then he was known as a patient, not a business, in the field.

Where the hype, questionable timelines and lack of detail do seem to be impacting, though, is in misleading the public as to what is possible. The more Dugan talks about the soon-to-hit-the-market consumer brain-typing device, the more the general public may think that BCIs are less 'science fiction' and more 'science fact'. The more Elon Musk talks about Neuralink on popular blogs such as *Wait But Why* and comedy podcasts such as *The Joe Rogan Experience* (this is how he seems to be releasing pertinent information about his company), without the product being unveiled, the sooner 'the boy who cried wolf'-type reactions to new discoveries and great science will appear: 'I thought we'd already done that?' Politicians and funders are regular people who read and watch the same things you do, so what's to say that reaction won't turn into a reduction in research funding that is geared towards people with medical conditions as opposed to consumers who just want some new, cutting-edge device?

Getting everyone more excited about BCIs is good as long as it isn't at the expense of helping those who need it. Yes, it's great to develop cool new ways of playing video games with your mind, or communicating with your friends, but it would be better if we could take any learnings or new technology in that space and apply them to the health sector, if it means life goes from bad to better for someone (as opposed to good to better for a gamer).

Public or Private?

These companies and internal teams are not just built by charlatans. They have good scientists on their books, who are experienced in the industry and keen to create technology that can reach beyond the tech-transfer issues of science spinouts and the politics of academia. The Silicon Valley companies have long-term visions, despite the bombastic headlines they garner about consumers having the devices in their hands (or rather, their heads) within the next few years. With the scientists they have on board and the approaches taken, from what we can tell, the companies know that it will be a long time before something concrete hits the market. And it is great to see companies set up with such long timeframes that are typical of research, but without the pressure to publish and report short-term incremental results to grant funders that comes with university-based discovery. Essentially, they can just get on with it.

'Just getting on with it', though, can lead to issues of transparency and accountability. Without the obligation to report to anyone outside their executive teams, the companies pouring money into BCIs are left to their own devices, as well as their own sense of ethics and consumer privacy.

I don't need to go over the issues unearthed about Facebook from around 2016 onwards, but there are understandable concerns when private, wealthy, largely non-diverse companies start creating technologies that underpin society as a whole. And when I first read about Musk's 'consensual telepathy' concept, I couldn't help but immediately think of all the 'consensual' activities of corporates over the evolving years of the internet that turned out to be quite non-consensual after all. Will we need a General Data Protection Regulation for the brain, for example?

There's an argument that BCI research should stay in publicly funded institutions, but that line of reasoning falters when you remember who is behind much of the government funding: the military.

BCIs for Soldiers

In 2018, DARPA announced its Next-Generation Nonsurgical Neurotechnology (N³) programme, which would be looking for viable non-surgical technologies that could eventually be used in healthy subjects for 'human–machine interactions with unmanned aerial vehicles, active cyber defense systems, or other properly instrumented Department of Defense systems'.[43] Dr Al Emondi, programme manager in DARPA's Biological Technologies Office (BTO), also said: 'DARPA created N[3] to pursue a path to a safe, portable neural interface system capable of reading from and writing to multiple points in the brain at once.'

It was clear that DARPA was looking to put this money into non-invasive BCIs, capable of both reading from and writing to the brains of soldiers, and in 2019, it announced the six research teams who would be awarded grants. Instead of investing in microelectrodes, the teams worked on acoustic signals, electromagnetic waves, genetically enhanced neurons and infrared beams, and at the end of the $120 million four-year programme they are to demonstrate their technology in a 'Department of Defense-relevant application'.[44]

DARPA's programme is under the guidance of external bioethics advisors, but there's an argument to be made that the military funding huge swathes of research in the BCI space is problematic. It's very exciting when news comes out about neurotechnologies that can augment our perception or lets us communicate brain to brain, but when put through the lens of military activity, suddenly the technology takes on a darker shade.

The idea of having private companies investing large sums of money in the field is a lot easier to stomach.

LETTING OTHERS INTO OUR BRAINS

There are many ethical questions around what might happen if we could effectively decode what goes on in each of our heads. For instance, what does that mean for countries with much more emphasis on surveillance? In 2018, there were already reports that train drivers, factory workers, soldiers and employees of Chinese state-owned companies were being made to wear EEG headsets to track mental spikes that might align with depression, anxiety or rage.[45] The explanation from the companies is that the headsets are about keeping up with workers' moods so they could adapt breaks accordingly, or to help train new staff when they are in VR headsets simulating different scenarios, but the possibility of abuse of power is easy to see and most likely very difficult to regulate if the headsets are allowed in those settings.

Many of those in the industry say it's still so far off that the panic and fear are unfounded. Headlines about companies and criminals hacking into people's brains and medical devices seem to be overhyped due to the nascent nature of the technology. Arguably, Alexa and Siri are far more invasive with their continuous underlying microphone, which listens to, records and doesn't allow you to delete your family's every word, and sends it back to their servers for the companies' own analysis and ownership.[46]

Saying something is far off, and therefore does not need to be feared or considered, is short-sighted, though. At what point do we start worrying? At what point do we start questioning? At what point do our fears and concerns and genuine queries get addressed, even if just to allay our concerns?

IMAGINING WHAT'S NEXT, NOW

And so there's another question to ask of brain–computer interfaces: when will having your very own BCI be a prerequisite to living on twenty-second-century planet Earth?

When researchers are applying for funding grants for BCIs, one of the biggest pushbacks they get surrounds the idea of the devices being implantable. In the funders' eyes, we need to get to a point where these things aren't implantable, in order to get more people on board. The key question here is: what makes an invasive device acceptable to the public? How do we measure sentiment around what is seen as acceptable when it comes to 'body editing' for the sake of goals: medical, cosmetic or simply a little improvement?

Think about it: cosmetic surgery isn't seen as too crazy nowadays. Yes, there are different levels of what is acceptable: boob jobs seem extreme in some cultures, dental veneers not so much, but certainly neither would be put in the same bucket as cyborgs. When you consider the whole space of female reproductive methods – implants, coils and pills – for some, these augmentations are for medical reasons, for others it's for lifestyle. When it's not done for life-saving measures, it's still accepted by society – in other words, the sacrifices that come with these edits or augmentations are seen as 'worth it'.

So what is it that will make BCIs 'normal'?

We have to remember that it was porn websites that took the internet from the lab to society. It wasn't the military or the scientists who built it that spun it out for mass use by broadcasting the success of online file-sharing; it was when entrepreneurial (or nefarious, depending on the legality of the porn) people who worked out a society-wide use-case that brought it to life, prompted other creative ideas (shopping, friend-connecting, news-reading, and so on), and made it 'normal'.

And the thing about the internet is that it's not just 'normal' in society: it's necessary. If you want to take part in the everyday ups and downs of living in an advanced twenty-first-century society, you need the internet. You need it for government services, you need it for information gathering, you need it for applying for jobs. You can't really opt out of having an email if you want to take part in the global economy, no matter how much you hate the idea of being online.

So at what point will BCIs be 'required'? At what point will sitting exams mean needing to enhance your brain (because the average score has been brought up by all those others enhancing already) or accepting that you'll not be able to compete with your brain-enhanced peers? At what point will medical provisions for those with Parkinson's start to dwindle because there's now a 'cure' for those who use BCIs, and those who do not or cannot are left behind? At what point will the gap between the haves and have-nots mean those who can afford one of Musk's neural laces and those who cannot, further widening socioeconomic disparities worldwide?

At what point will opting in to BCIs not feel like a choice for active members of society but, rather, a requirement?

When Elon Musk was interviewed by Axios about Neuralink, he said: 'How do we ensure that the future constitutes the sum of the will of humanity?'[47]

He was referring to his belief in the future conflict between humans and AI, but surely creating a device that could create further wealth divides, distract from helping those who need it and satisfy the wants of the techno-fanatical few is in no way the sum of the will of humanity, but rather the atomisation of it.

WHY DO WE EVEN WANT THIS STUFF?

BCIs make complete sense for patients. For those who have become paralysed, or are born without certain senses or have minds that start degrading with neurological diseases, BCIs offer life-changing alternatives to having to live with a condition.

For the rest of us, though, considering the societal costs that may come with a consumer BCI future, the question remains: why do we even want this stuff? Are we so obsessed with productivity and control and being better, faster, stronger as human beings that we cannot see the downsides to giving up access to our brains? Are we so taken by the cult of the entrepreneur that we listen and deify without questioning inherent motive and privilege in their actions? Are we so impressed by the utopian worlds of science fiction, thus distracted from the similar, more dystopian tales, that we simply *must* have those devices of the future?

Yes, I want that headjack from *The Matrix* so I can learn every skill and read every book. Yes, I want that 'point-of-view gun' so I can be better understood and heard.

But do I *need* them at the cost they come at?

The business world isn't there to ask those questions. The market supplies what is demanded by those buying: us. The sentiment of 'we're just giving the people what they want' is a common excuse for problematic behaviour.

We cannot be blinded by our desires, and 'how cool' something would be; we must consider the other side of the story: how uncool it might be for someone else, or the cost of our unchecked desires on how society evolves.

In that same Axios interview, Musk said: 'We're like children in a playground . . . We're not paying attention.'[48]

I couldn't agree more.

Taking Responsibility for Artificial Intelligence

Have you seen those Boston Dynamics robot videos? Where Spot, the robot 'dog', or Atlas, the robot 'person', is climbing up stairs or opening a door or leaping between boxes laid out across an obstacle course? With every new video you see – there's a new one every few months or so – the robots demonstrate their ever-more flowing movements and put their mechanical muscles on show. The way their metal bodies move reminds me both of dystopian AI films as well as impressive sports people competing in *Ninja Warrior*. They are increasingly and eerily life-like with every iteration.

What are not always shown in the videos are the robot handlers – the scientists who are always right next to the robots, just slightly off camera, with their remote controls and their focused expressions and their looks of delight when their masterpieces do as they hoped. I say 'hoped', but really I mean 'programmed' – the robots are essentially choreographed in code; they aren't doing these tricks of their own volition.

Sometimes there are calls to celebrate those researchers – to remind us that these incredible feats are the work of real people, with faces and names that don't make it onto the videos. What's followed those calls are sometimes responses along the lines of 'Wait, the robots aren't entirely autonomous . . .?'

The extent of robot puppetry is not the point – the technology is only going to get better over time, and require less and less real-time input from the robot handlers on the side. The point is that we're already forgetting the fact that humans are the ones in control. They are the ones building these robots – dreaming them up, programming their moves before and during the demos, deciding on the systems being used to train the algorithms. Regardless of how good the robots get, we must never forget those who made them.

Celebration of achievement is one thing, yes. More importantly, though, is responsibility. If we don't know who we're meant to congratulate, how are we going to know who to blame?

The thing about AI is that it's not all that it's hyped up to be. It's not the crazy *Terminator*-esque robots, and it's not HAL – not yet, for sure, and maybe never. Even if we can build technology good enough to make all kinds of decisions for us, really it's about whether humans continue actively to choose to use it. AI is a tool that humans decide to put to use, so instead the questions must surround whether we continue to keep incorrectly thinking that it's happening outside our control – keeping our responsibility blinkers on to save us having to face up to hard decisions – or start to understand what's happening right in front of us: we're choosing to use AI.

To understand the lengths to which humans do have a 'say' in what happens with AI, it's key to understand exactly where humans have control and where that control starts to waver. To do that, we need to know how AI is built and, thus, where the human input and decision-making really happen.

QUESTIONING THE DATA
The first thing we have control of is the data.

Simply put, AI works by taking a large amount of data and having a computer sift through that data and find patterns within it, to then take some kind of action. An example might be data concerning what clothes you buy online. Maybe there will be a list of the previous purchases (and returns), the pages browsed, the 'favourite' items, and maybe even more generalised information such as 'what's popular' in your country, what age you are and so on. AI might be used to comb through all that data – that information about you – and then send you an email once a week with the 'best matches' for you, to better incentivise you to spend money (as you, in theory, will be shown items that you like and will find harder to say no to).

The dataset that the AI uses is built, and decided upon, by humans.

The people on the front lines of the AI industry are the data scientists. Most of a data scientist's job is what's called 'data wrangling'.

Imagine you wanted to create a program that searches the internet for all the media mentions of companies around the world. First, you'd gather a list of all companies so you'd have a master database to compare to, and then track the mentions. To get that complete list, maybe you'd go to each country and request a list of all the registered corporations from the governments. Each government probably has a different system by which you request the data – maybe some have an online portal, maybe others require the request to be put in writing – and eventually you end up with almost two hundred documents with lists of companies inside. Maybe some of the documents are Excel spreadsheets, maybe others are Word documents. Maybe some documents have 'Ltd' or 'LLC' after the company names, maybe others simply put the brand name. Maybe some documents have the data in a vertical list, maybe others have them listed with commas in between. Maybe some countries define 'companies' differently to others (do they include B Corporations and social enter- prises too, for example?). If you wanted to create one document of your own with all the companies in the world in it, it would take some time to check through, clean up and compile all these different country docu- ments into one.

Only once you've cleaned your list could you start building your program. You could build your code to define where to search online for these mentions, determine where to pull in those found results to, and then scan the articles for mentions of any one of the companies on your master list. Beyond this, though, you'd also have to ensure that the program you build takes into account mentions of companies with and without Ltd and LLC behind their names, or misspellings, or differing capitalisation rules from different media publications, and so on. If you want to be able to trust your program, you need to make sure your data is robust. This process, as you can imagine, takes up a huge chunk of data-scientist time and can be prone to error and affected by the bias of the person tasked with the clean-up job.

Think about compiling all the health records of everyone living in the UK, and all the different doctors and nurses and administrators who enter information about each person, and the chances of all of them entering the data in the same way, error free. Or think about a data scientist who maybe doesn't understand the nuance of a particular industry and deletes parts of datasets in a bid to clean them, without realising how crucial those parts of information are in the real world (or makes connections between various sources that shouldn't be connected). The AI system doesn't know anything beyond the dataset chosen and cleaned by the data scientist; it works off this as a 'true' reflection of the world, regardless of how true it really is. AI is only as good as the data on which it is built and used to train it, and decisions around data are ultimately made by sometimes-fallible humans.

There are many questions surrounding data, which we, as the public, must ask of those building AI systems. There are questions around how we collect data: are we collecting it fairly, with the consent of the individual, and without bias? There are questions around the ownership of the data, and what the owner chooses to do with it: Is the owner the person that the data relates to, or the person who took the trouble to collect it? There's the regulation of the data (which countries choose to do what, and for whom and when?), the usage of the data (who gets to use it, for what and when?), and even the value of the data (how much is it worth, and to whom? Is having an app which monitors heart rate more valuable to the patient or to the pharma company that now has a huge dataset of a population's heart rates over time, and can use it to design, build and sell its next product to a huge market, making billions of dollars? Can it be both useful and economically exploitative?). When these questions aren't asked, we run the risk of having AI systems which haven't really been properly checked at the root of society's digital products and services.

*   *   *

The datasets that power our world (in the sense that they are the ones we use to build systems upon – AI-powered or not) are also biased. There are the datasets that determine the safety measures for cars, that don't account for women's measurements, making women 47 per cent more likely to be seriously injured in a car accident and 17 per cent more likely to die.[1] There are the studies that determine the approval of drugs for respiratory ailments, which African Americans are more likely to suffer from, of which only 1.9 per cent included minority subjects (meaning the drugs are tested on and found to be safe for almost exclusively white people).[2]

But how do we create universal datasets to base certain AI algorithms on which are both unbiased and also reflective of society? The example that is regularly cited in discussions of biased data at the root of AI systems points to Google Translate: if you type in a phrase such as 'He is a nurse' or 'He is cooking' in English, translate into a gender-neutral language such as Turkish and then back into English, it changes to 'She is a nurse' and 'She is cooking'.[3] Of course, in this example – in English – it could have returned 'One is a nurse' or 'They are cooking' but, beyond those semantics, we have to also appreciate that the world is configured in such a way that women (rightly or wrongly) *are* more likely to cook or work as a nurse, so the AI translation algorithm doesn't necessarily have the wrong idea about the world, but rather the wrong idea about the cost of the error or assumption (not that algorithms have ideas or assumptions, but you catch my drift).

At the end of the day, society right now *is* biased – women are paid less, ethnic minorities are more likely to be imprisoned, those of higher socioeconomic status are more likely to be politicians and CEOs, the list goes on. If we are building algorithms that model and control the world, then they need to be reflective of the world as it is in order to then 'make it better' or more organised, efficient, fair or whatever else the algorithm is set up to do. But do we create a dataset that is more of a 'perfect vision' of society, and almost 'trick' the algorithm into building on top of what we agree to be the society we want (so that the algorithm doesn't accentuate societal flaws)? Or do we reflect society as it is now and ensure that whenever we create an algorithm, we place plenty of 'red flags' within it to ensure the systems know that we're not happy about the inequality within the set, and to make sure not to accentuate that? Do we trust the builders of the algorithm to do that (a lot of software is proprietary and can't be externally reviewed)? After all, what really *is* a reflective version of society at different scales and, even more puzzling, what do we actually all agree is a 'fair' perfect vision of society? And who decides?

The thing about databases, though, is that they're not that sexy to talk about (at least for the majority of us) and no one really wants to take responsibility for them due to the complexity inherent in getting them right, but they're key to focus on – more so than the algorithms – when we talk about AI and the effect it has on society. When we talk about AI, we need to focus on these questions of use, misuse and analysis of data, and the social systems around that, even if they aren't quite as interesting as tales about *Terminator* or HAL, and ultimately give us even more to think about and add to our to-do list. We cannot succumb to boredom, laziness or defeatism when it comes to questioning those who build the systems that our future societies will run on.

CONTROLLING THE METHODS OR LOSING CONTROL?

If data is the first thing we can control, the second thing we can dictate is the type of methodology we use to analyse that data and what problem we point the AI system at.

There are many different ways AI systems are built. There are the more dictated step-by-step processes such as decision trees and linear algorithms, where those building them will decide exactly what the algorithm is to do and in what order. There is unsupervised machine learning, which is basically where a computer program is set loose on a load of data without being told what to look for and for what purpose, and it finds its own patterns. The benefit is twofold – the computer learns how to learn itself (in this specific context), without needing further input from the human, and new patterns are found which humans didn't already know to look for. We can use AI systems for all manner of things. There is natural language processing, which is the area of AI dedicated to reading, understanding and generating human language and speech.

There is computer vision, the area dedicated to having computers recognise what's in a photo or video, in real time or not, and then maybe taking some form of action in response.

The type of methodology used, and the 'thing' the AI system is pointed at, is built, and decided upon, by humans.

Right now, in 2020, there are limits to the power of AI. We haven't got to the stage yet that we can 'point' AI at any kind of problem we want solved, and it can go off and start gathering insights and acting on its own – it still requires huge amounts of human puppetry to get it to do the tasks we desire. We also haven't yet got to the stage that AI can randomly go outside the bounds of what humans design it to do and effectively succeed in creating something that impacts society. For instance, if we task an AI system with calculating projected profit of a company, it won't go off and rewrite the constitution and start implementing and enforcing new laws on humans. It can find weird, wonderful and unexpected ways to answer that profit question we pose it, but it still only 'knows' how to stay within the realm of the data and methods attributed to solving that problem, and that problem alone.

Humans have control over what types of algorithms are used, and exactly what they're used for, but where we start to lose some level of control is in knowing how exactly the system goes from the 'input' of our question to the 'output' answer that it presents. And it's here – where our control starts to waver – that we must consider our active decision to use AI systems to come up with answers for us.

BLACK BOXES

AI algorithms can be built in many different ways – using different methods and different datasets at the core. These different approaches result in different algorithms, which differ in how easily you can interpret the way they get to the solution they present.

For example, think of a machine that helps you decide what book you should read next. You put in all the books that are on your bookshelves, and you get out one single book the machine deems best to dive into next. If you were to build the machine using basic methods, you'd be able to open the doors of the machine and see exactly how it calculates the best book – maybe it simply orders the books by how long you've owned them and serves you up the oldest, or maybe it googles the most popular book out of all the ones you own and serves you up the one with the best Amazon reviews. It would be easy for you to work out how the machine made the decision.

If you decided the age of the book or the most highly rated on Amazon wasn't a very good indicator of what was going to satisfy your thirst for reading, you might build the machine using more complex methods. Maybe you'd want the machine to take into account your previous ten books, or analyse your current mood, or compare your reading habits to that of the most successful people on Earth and adjust accordingly. Maybe you'd want the machine to read every book and pull out the one that varied the most from your most recent read, or predict the next stock market bubble and pick the book that best prepares you for understanding the future markets. Maybe you would want a combination of all of the above, and more. If you wanted to build a complex machine like this, you might employ a class of AI algorithms that have what are called 'hidden layers'. One such example is what's called a 'neural network' – modelled roughly on neurons in our brains – and these hidden layers play the role of clustering and classifying the data you input, analysing it, and answering the initial question posed. If you were to open up your more complex machine, you wouldn't be able to trace exactly what the machine was doing with the book options, or in what order – you'd simply know what went in and see the result of what comes out.

These more complex algorithms that cannot be interpreted easily are called black-box algorithms.

Sometimes the unwanted result of a black-box algorithm isn't a problem: if the machine happens to get the book recommendation wrong once or twice, it's not hugely important in the grand scheme of things. But it is problematic when AI systems determining human fate come into play – systems that decide how expensive your insurance should be, or whether you should go to prison. This is because contesting those decisions is nigh-on impossible if we don't know exactly how and why the decision was made. And if a mistake should indeed be made, working out who should be held responsible for that decision becomes complex.

In New York, March 2015, nineteen-year-old Frank Thomas was convicted of criminal possession of a weapon, reckless endangerment and menacing a police officer; and was sent to prison for fifteen-and-a-half years.

The previous August, police had pulled over a car for driving without headlights, the driver and passenger ran away into a park, and as they were chased, the police heard a gunshot. The police never caught them, but Thomas was the owner of the abandoned car and the prosecution found his DNA to be a 'likely match' to that found on the gun. The DNA analysis was the only physical evidence that linked Thomas to the gun, and Thomas maintained his innocence.[4]

The gun was found to have a mixture of at least four people's DNA on it – maybe even five or six – and to assess the likelihood of Thomas's match, private company Cybergenetics was tasked by the prosecution to use its AI-powered TrueAllele DNA mixture interpretation product to link the defendant to the gun.[5] It's widely accepted by the scientific community that complex mixtures of DNA when tested are unreliable when tested the 'traditional' way. There's a lack of accepted scientific validation of Cybergenetics' ability to match DNA when there are more than three people's DNA in the mix, and yet its analysis put Thomas in prison. (It has a level of validation for fewer than three people in the mix of DNA through peer-reviewed scientific papers, but it's still controversial as the tests have all been conducted by the company itself, and the source code is proprietary and therefore unchecked.)

The company maintains that its source code shouldn't need to be scrutinised externally as that would put the profitability of the company at risk – so a black box, possibly in the sense of an uncheckable AI system, but more crucially as code simply being hidden by a private enterprise, is essentially being used to put people in prison without true accountability or validation. According to Cybergenetics' CEO Mark Perlin, the company is going for the 'complete removal of the human being from doing any subjective decision making', which – considering the questions hanging over Perlin's refusal to properly validate his company's technology – is worrying when it relates to putting people in prison for years or even life.[6]

Black boxes are common in the AI field – both in the sense of the uncheckable algorithm as well as the push to keep software proprietary – and it seems there is already plenty of controversy relating to current technologies out there in the justice system. Another algorithmic DNA-testing tool, built by the Office of the Chief Medical Examiner in the US – the Forensic Statistical Tool (FST) – was in fact blocked from being used in a 2015 case due to its black-box nature. It was then released in 2016 to a third party to check its validity and was found to have errors in the source code – the expert witness reported that the program excluded potentially important data from its calculations, which could 'unpredictably affect the likelihood assigned to the defendant's DNA being in the mixture'.[7]

In short, both kinds of black boxes – the technical and the proprietary – are a complex reality of the AI-powered algorithm space. And when you consider circumstances where human lives are at stake, the prospect of not being able to take a look inside the decision-making process is understandably freaking people out.

There are many proposals and ideas to combat the issues black-box algorithms present in the AI-powered world of which we're increasingly becoming a part.

Researchers are working on all sorts of measures to make black-box algorithms more interpretable.[8] For example, to work out why an algorithm came up with some kind of answer, they might probe the system with different kinds of questions to spot the 'rationale'. They might pick different segments of the data to run the system on – segments that are both representative of the whole, as well as ones that are more like outliers – and see how the final answer shifts. Researchers can sometimes also work out not exactly how the algorithm works, but which parts of the data most contributed to the final decision, and therefore find which pieces of information are more or less important to the system (for example, data about black people or white people, or men versus women, and so on).[9]

Other researchers go down the regulation route, proposing new rights for citizens for whom AI systems have decided their fate. The 'Right to Reasonable Inferences', for example, proposes that those who control which data is used in an AI system must be able to justify exactly why their chosen data is indeed a reasonable basis for the conclusion reached.[10] In other words, it would force the system builders to prove why their dataset is representative of the norm, why any connections made in the data are fair and relevant regarding whatever the system is used for, and how statistically accurate their system is. When it comes to contesting decisions that might have been made on the basis of race or sex, for example, this Right to Reasonable Inference might be crucial in exposing bias.

There's much discussion around the idea of machines making indecipherable decisions concerning humans – and rightly so. But sometimes in those discussions we miss the point that black boxes are not unique to AI systems – and to understand this, we first have to take a little detour into the wonderful world of moral philosophy.

DINNER PARTY PHILOSOPHY

There are a few popular moral questions that tend to be brought up in AI discussions – the most famous being the 'trolley problem', a thought experiment in ethics. It goes like this: a trolley (or a train) is going along the track, heading straight for five people standing on the rails (they cannot move for whatever reason). It's going too fast to stop, but there is a lever at the side of the track that would switch the trolley onto a different set of rails, where only one person is standing, and who also cannot move. You are standing by the lever and have two choices – don't touch the lever, leaving the trolley on the tracks it's currently on, killing five people, or switch the tracks, and kill one person.

You could say that the utilitarian thing to do would be to switch the tracks, saving five and losing one – 'better for the many'. You could also say that if you partake in the switch, you personally become responsible for the death of the one person, whereas if you don't touch the lever, you can't be held accountable. But then there's also the 'moral obligation' idea that, if you're there and watching, you're a part of it, and it would be immoral to not touch the lever if you value five lives more than one. But then you can't compare five lives with one, as it means thinking about how to value a human life in the first place (if the five people are elderly and close to death versus the one person being a cancer scientist 'close to a breakthrough' . . . or if the five are mothers of young children versus the one being the cancer scientist – how do you compare these?).

The problem is often touted as relevant in the development of self-driving cars: how does the AI choose a course of action when faced with options that all result in human death? Does the system find a way of making so-called 'cold logic' predictions on human life? (Much like in the film *I, Robot*, based on an Isaac Asimov short story, where the main character is rescued from the car that was sinking in the river, rather than a twelve-year-old girl, as the robot found her to be less likely to survive the rescue.) Or does it take human productivity into account – considering what effect on society each person has and prioritising the more 'useful' of us? Or maybe it prioritises the person in the car, treating them as the owner it should be 'loyal' to.

With the acceleration of technologies such as AI, some say we are forced to come up with some kind of solution to these moral problems – that we need to 'fill the gap' left by those who posed but didn't answer them. The argument goes that the philosophers who first came up with these ideas didn't foresee a world where we really needed a solution at scale: if the scenario posed in the trolley problem ever happened in the real world – there really was that one individual at the side of the track having to make that godawful decision – this awkward situation would be an unlikely and unfortunate occurrence. They didn't expect societies of the future to create cars and lorries and all sorts, and require this kind of morbid decision-making capacity every day. This argument proclaims that with this need for an answer now crucial, we need to stop leaving these problems 'unsolved' and up to each individual's moral standpoint, but rather treat them as problems that we as society have to solve together, now.

We can easily get caught up in trying to solve these moral riddles. The panels at technology conferences trying to 'get to the bottom' of the trolley problem, or the articles posing yet another moral riddle for those reading to ponder on, are only as useful and interesting as a dinner-party conversation. They are fun riddles, if a little morbid, for us to occupy ourselves with, but not at all useful in bringing the public's attention to the unanswered questions surrounding those building the technology. Someone has to make decisions on data and algorithmic methods, or has to give the go-ahead for uninterpretable technology to be used to make decisions about human life; whether that AI system has solutions around the trolley problem in-built or not.

These moral riddles keep us at the 'big picture' level, and distract our thoughts away from real-world questions such as how to regulate the development of this technology, or who is the one building the systems, or what datasets they are using, or querying if predictions of human value are included in the system (and if so, what sort of results it is generating).

These riddles also paint the technology as more unhinged and independent than it is – like it's some kind of zoo animal let loose on society, that will make moral (or arbitrary) decisions we're not OK with, and as such shouldn't be let out into the world.

What we fail to acknowledge here is that we already deal with similar 'moral conundrums' in everyday life, and have done for years. Take 'rationing of care' for example – where a healthcare provider refuses to give treatment to a person who is demanding it.[11] When someone is close to death, the ethics of the situation are complex. Going against patient autonomy (their right to make their own clinical decisions) is not ideal, but sometimes medical professionals will deem the treatment futile and their own role would be compromised, in that they are being asked to use their skills in an inappropriate manner (the treatment may not work, it might harm the patient more than help them or result in an undignified death). In the case of public health systems – where the taxpayer, not the individual patient, is paying for the treatment – the cost of a treatment that is deemed futile (in terms of monetary cost as well as cost of not affording to help another patient instead) also comes into play.[12] The decision to not give someone treatment on funding grounds – which is common in public health systems – makes sense from a utilitarian standpoint, and of course directly relates to the trolley problem.

The point is that these ethical conundrums, which are often the focus of public debate around AI, are not new. That doesn't necessarily make them any less interesting, but they are a distraction when the other questions and considerations of law, responsibility and regulation aren't also discussed (and acted upon) to the same extent.

THE BLACK BOXES WITH WHICH WE'RE ALREADY LIVING

We already have so many black-box algorithms we live with and accept in society. Much like rationing of care, we might know that they are problematic and complain about them from time to time, but we are in no way new to the idea of an impenetrable black box of decision-making, rationale, bureaucracy and layman-expert tension we cannot always access when we want fully to understand the result of an organisation's decision.

There's a famous series of sketches in the UK television show *Little Britain* that centres on a receptionist character in various different offices who, in each sketch, deals with some form of very simple and obvious request from someone not by using common sense, but by tapping mindlessly on a computer, asking mindless questions and eventually always coming to the conclusion: 'Computer says no.' She always mimics both an incomprehensible checklist questionnaire irrelevant to the task at hand, as well as a clunky website which – for the life of you – you can't get to just bloody well work. They are famous sketches because they ring true to anyone who's tried to navigate any kind of bureaucratic system, either online or off.

There's a culture of following rules, no matter how daft, without questioning them (and assuming they are there for a reason) throughout society – both in terms of those working within a system and those coming up against it. Sometimes we overly trust systems – computer-based or not – instead of following our own common sense. At Georgia Tech in 2016, researchers proved that humans were more likely to trust a robot (with a sign on it reading 'emergency guide robot') pointing down an obscure corridor in the event of a fire alarm, as opposed to following the emergency exit signs clearly pointing the other way, from which they entered the building.[13] There's inertia in 'standing up' to perceived societal rules and 'expert' figures, in trying to find a solution around them when need be. That inertia exists when it's easy to see how to go around the computer program; it's even more apparent when it's tough (but clearly required) to find a better solution. An example might be getting around a complex healthcare computer system that you've been told by your boss to follow, and which won't let you sign in a pregnant woman about to give birth, versus calling up any maternity doctor right away and contending with the computer system (and possibly a boss's wrath) later.

It's easier to 'blame' the system as opposed to being creative in finding a solution. Anyone who has had to call up their local council to ask for some small but obviously vital request – such as for them to remove the illegally dumped mattress blocking your building's entrance – you'll know exactly what *Little Britain* was trying to get at. We've been dealing with (or not dealing with – and getting grumpy about) opaque systems and the gatekeepers of action forever.

This black-box problem is not unique to the current 'age of AI', it's part of our society. AI is not this separate thing we don't control and which independently came into being – it's an accentuation of humans and the society in which we live, which we build and affect and can change, if we desire. If we are worried about the effect AI might have on society, particularly with ethical decision-making, we need to accept that this worry is really about the actions of other people, not machines. This acceptance does two things. First, it stops us from being scared of AI and rather points our worry and therefore our blame and expectation towards the humans building the systems and making the decisions to use these systems despite the societal concerns. Second, it shines a light on the 'black boxes' already common in society that we've tolerated over the years. It makes us (hopefully) rethink the indecipherable and therefore unacceptable black boxes that slow progress, making people at the very least frustrated and at worst put in harm's way, which are at the root of many powerful businesses and governments today. A 2016 study found that one of the biggest forms of negative impact for those in the UK who wrongly lost their disability benefits – beyond losing the money they needed to live on – was coping with the never-ending cycle of bureaucracy when they tried to appeal the mistake.[14]

By questioning decision-making in AI systems, we must also question the design of society itself. The optimist in me also says that if we're to make those changes to society first, then maybe the AI systems we build will have a better functioning society as opposed to a clunky one as its foundation – 'we should probably change that before the AI thinks that's actually what we want . . .', as it were.

The ethical considerations and consequences of AI, or the problematic outcomes that might come from the use of it, are wrapped up in the ethics of the organisation it's emanating from and the society in which it's embedded.

Even more so, they are wrapped up in the ethics of individual human beings – ourselves.

Our brains are 'black boxes' – we regularly make decisions and don't know exactly how or why we got to the conclusion. Maybe AI systems are less fallible than our own brains; in 2015, a study found AI systems to be better than humans at hiring – when the algorithm picked the candidate, they stayed at the company longer than those chosen by humans.[15] But then maybe we're also just wanting to be rid of the responsibility that comes with making hard decisions.

Humans use a lot of probability theory and don't realise it – there's a lot of intuition in making life or death decisions. Some decisions can feel instinctive, but when it comes to complex problems, a lot of decision-making has to do with experience and memory; in other words, the database of information on which our system is built. We draw on emotional moments for 'advice', we access our memory to process what happened before and therefore what might happen next – there has to be a formula for how we access and combine these memories to come up with an answer, surely, but we don't know what it is and yet we trust it.

The job of the historian is to go into society's long-term memory, analyse it, pick it apart, and find out what's missing or buried, to then help society make better future decisions – though, of course, many times we don't learn from history and repeat mistakes. Maybe an AI system wouldn't be so 'forgetful'?

WHO'S IN CHARGE?

People aren't all bad. Humans are pretty impressive beings, and when we harness the power of technology and augment – instead of replace – we can become super-powerful, and sometimes, super-good.

We don't create machines to be our overlords, we create them to be our assistants. You only have to look at Siri and Alexa and the other assistant technologies to see the demand for tech to do our dirty work for us, and help us navigate this complex thing called life. IBM has called its AI technology 'Watson' – and despite the many discussions about Dr Watson actually being the brains behind the operations in the famous Arthur Conan Doyle suite, we all know it is Sherlock Holmes who is the master of the case, with Watson by his side. Sherlock might have been given a piece of information or a hint from Watson, but he's the one driving the thing. It's not Watson that's stealing your job, it's Sherlock telling him to.

But who is Sherlock? What's Sherlock's story? What's Sherlock trying to achieve? We know he's trying to solve the case – yes – but part of the story of Sherlock Holmes involves his nefarious means of getting what he wants. In some interpretations of the Sherlock Holmes stories, he breaks the law and makes short-term immoral decisions (for the good of the long-term case). He does things that are below the belt in the name of doing what's right – and seems to have a clear view of what he thinks is right and wrong, not caring if we understand it or not.

The difference between Sherlock and Watson versus us and our AI is that they are only two people affecting those around them. We and our AI can affect the whole world and, of course, we're not talking about a work of fiction, we're talking about the real world . . .

Sherlock is also just one person – he can be held accountable for his moral judgement in his fictional universe. With society-wide use of AI, sometimes it seems we need to come to some form of consensus as to what's morally correct and just – a nigh-on impossible task.

When we get caught up in thinking that the moral considerations that AI faces are unique to the technology, we run the risk of disassociating from the responsibility of solving them. We think of the problems as too hard to solve, too difficult to regulate against, impossible to properly consider until the systems are already in the market (at which point it would surely be too late). We accept the excuses from companies when they say they can't show us their workings as their profits would suffer from exposing proprietary information, and listen when they say they couldn't show us anyway because the AI does it itself.

When it comes to things potentially 'going wrong' with AI, it's one thing to tell ourselves kinder narratives around the difficulty of the situation, and let ourselves off the hook in some sense, it's another to move the blame away from ourselves altogether, and onto a piece of machinery. We need to remember that we are Sherlock, and that AI is Watson – only assisting, not driving, and never the one to blame.

Words Matter

One of the most popular phrases and lines of thinking when it comes to AI is 'robots are stealing our jobs'. It's a compelling image – the *Terminator*-like humanoid coming along with its high-tech guns and mechanical muscles, and taking what's rightfully ours as human beings ('You can take our lives but you can't take our banal admin jobs!') – but the reality is, frankly, quite different. Jobs aren't being 'stolen' by robots; companies are making active and deliberate decisions around saving money on human labour and assigning that money to paying for a technological upgrade. Now, of course, as we build ever-more intelligent machines, and we manufacture robots in such a way that they save businesses exponentially more time and money on arguably quite boring, monotonous and manual work, of course we should use the best tech out there to do it cheaper, better, faster. It's not just about companies saving time and money, after all, it's about societal productivity going up – more food, more cars, less waste and so on – it's a pretty utopian view of what our world could look like, but not an unreasonable one.

But people could be made redundant and as such not be able to pay to live in this new utopian society. There are 75 million people around the world who work in jobs ripe for automation, likely to have been replaced by robots by 2030.[16] The World Economic Forum reckons there'll be 133 million jobs created by the same date, due to new businesses being created as a result of technological development, as well as socioeconomic growth. But there's the question of how those jobs will be filled if those 75 million people aren't effectively trained up in new skills. Our societal economic systems aren't yet built for abundance of goods and for people not needing to work – they are built on scarcity and the law of supply and demand – and so we don't yet know how to build welfare, health and financial systems for healthier people living longer without traditional sources of income. There's plenty out there that you can read about Universal Basic Income, 'taxing robots' and all sorts of other proposals for this future world AI systems are paving the way towards, but the point I'd like to get back to is the simple use of hyped-up but misleading language.

Robots aren't stealing any jobs – corporates are making active decisions to put people out of work and not take on the responsibility of retraining them. I'm not here to say that automation is bad, but I do think by continuing to blame the technology, we're increasingly letting companies quietly get away with putting profits over people.

Similarly, when we say we want to 'tax robots' (an idea that is backed by famous figures such as Bill Gates), what we're saying is that we should create an add-on tax for companies that choose to automate, in order to fund the loss of income for the human labour force replaced by said robots.[17] Whether that tax is used for training schemes or a Universal Basic Income pot, they would be covered. Language, and the way it can nudge us in particular ways, can be powerful in another way. If we shift the phrase to 'tax companies', as opposed to 'tax robots', that could easily turn into a narrative around taxing the 'most innovative companies' as opposed to the more traditional, immovable and stubborn ones. I can't really see that argument being as popular in the business world – even though the repercussions are exactly the same. (If you think that taxing big companies to cover the loss of jobs due to automation is a good thing, maybe you *should* in fact say 'tax robots' if you want it to happen without the loud criticism!)

Even when we're talking about AI in a more positive light, we use phrases such as 'AI will help cure cancer' or 'This government used AI to reduce their carbon footprint'. AI will do X, it'll do Y, and then it'll do Z – we read it in articles, we hear it on stage and we see it in company pitches. There is, of course, a benefit in talking like this: it simplifies the conversations around why AI is good (which are needed and true and important – AI systems really can be awesome forces for good, when done with care and responsibility for society at large). But it also externalises the actions being done. AI isn't the one speeding up cancer treatment or saving lives on the roads through predicting car accidents before they happen; it is humans building businesses and programs that are doing this for us, by choosing to use AI systems that help in those quests. It's the humans that are doing good things using the tool that is AI. They are doing good alongside the bad: humans are also not accounting for bias in algorithms and putting profit before jobs and incessantly serving us ads we didn't ask for.

ASSIGNING BLAME ALONG THE CHAIN

It's not a new thing – assigning blame and credit to technology as opposed to people. Look at the way the narrative 'smartphones have taken over our lives' has entered and taken over the public conversation.

The reality is that phone companies have designed them in such a way that we become addicted. They take inspiration from research in the fields of psychology, behavioural economics and game theory to work out what 'nudges' the human brain into different forms of actions. Whether that's keeping us coming back through the digital dopamine of notifications and likes, or having us follow the complete journey from seeing an ad to making a digital purchase, companies want and need us to keep using their technology in order to stay profitable.

We're so shocked by the idea that user-experience designers, software developers and all manner of other types of people in business are employing these methods for the purpose of getting us to do things we're not really choosing to do. They're not bad people – they normally don't realise the broader long-term impact (or rather, they simply don't think about it at that global scale beyond 'getting the person to click that button'). We don't want to blame these individuals as such, but we do want to be more in control of our technological addiction: think about how many articles there are out there on how to switch off notifications and turn your phone to greyscale and 'take back control' of your device.

We make a villain of the technology as it's too hard to compute how this stuff could be done by humans who care. It's too clunky and long and, dare I say it, pointless to know how exactly we got to the point that we as a society are addicted to little black mirrors, as it happened through a combination and a sequence of people making decisions not realising the impact of their small actions. We don't know at which 'point' in the chain the 'bad' decision or the negligence happened that led to our addiction – from the very first scientist to think up the idea of a computer in your pocket through to the behavioural science researcher who first looked into how to nudge people into doing things automatically, to the big companies creating en masse the smartphones and watches and apps and websites to the UX researchers adopting behavioural economics to make navigating a tech product easier.

We want it fixed, but it's too hard to work out who to blame and therefore who to hold responsible. The cost of paying for the systematic error (and by that I mean rebuilding our entire digital society, as so much now is built upon our usage of and addiction to our devices) is huge, and we don't know where to start – so we blame the computers and read our 'phone hack' articles in a bid to move on with our lives.

You can talk about this kind of 'chain of responsibility' across all areas of science and technology through the process of tech transfer. If you have a scientist researching something in a lab, which then gets spun out of the university into a business, which is then bought by a bigger business and sold on the market – which is then a bad thing to have in society (think: atomic bomb, subliminal messaging used in advertising, and so on) – who exactly is to blame? The researchers, who were simply trying to work out the answer to a question? The university, which was simply trying to pay off the money it invests in research by selling some of the results of said research? The initial business, which was simply trying to get good science into the market, before realising the only way to pay back its investors was to turn to a more financially fruitful but less socially kind application? Or the bigger business, which was simply addressing the demands of the market? It's hard to pinpoint – and sometimes there isn't an exact pinpoint; but by not doing the hard thinking at every step, pondering what negative way this development might go, and designing in such a way for that not to happen (as opposed to moving forward blindly), it goes unchecked and unblamed.

All of this is to say that in shifting the responsibility to the machines as opposed to the humans behind them – in using dehumanised language and not questioning the narratives we're fooling ourselves with – we are assuaging the guilt and systematically building a tougher system for future generations to contend with. They'll inherit the AI systems we build, and if we don't build responsibly, who knows what they'll end up with.

We cannot think of AI as this 'future thing' we might need to work out the trolley problem for. We need to think of it as a technology many people are building right now, and we – as a society – all play a role in ensuring what we build isn't 'accidentally bad'.

DECISIONS MADE EASY

Being more aware of what's going on behind the scenes in the AI field, and asking crucial questions about the decisions made by those building it as a result, are important. Beyond questioning databases, black boxes and company intent, there's another question which is often forgotten, which technology idealists and optimists tend to scoff at: do we even need technology to be doing this?

Companies and governments have lots of decisions that need to be made every day (in fact, every minute) – more decisions than any human can really take on. Those in the online space need to make decisions about what kind of results to show when someone searches 'what shall I eat tonight?' Those in the manufacturing space need to make decisions every second about whether a half-built car needs to be sorted left or right on a factory line. Those in the logistics space need to make decisions about which packages go out today in order to be delivered on time tomorrow or the next day. Those in the medical space need to make decisions about what drug to give each patient at which dosage and for how long. Algorithms and AI systems are allowing these organisations to keep up with the pace of demand of humans all over the world, by getting rid of decision-making that is banal, simple and 'just needs done' to get things from A to B.

All of this seems OK for 'simple' decisions but at what level of decision complexity or impact is it not OK to devolve responsibility to a machine? Which decisions arguably need to be done by a human in order to stay fair, true and nuanced without worrying about a computer 'getting it wrong'? Decisions such as which content to display and which to take down for being inappropriate or even illegal. Decisions around pricing of insurance, which may unfairly single out those at a disadvantage. Decisions around justice – putting someone in jail or not – which may unfairly use historical racist data as the benchmark.

The choice of using an algorithm instead of humans to do the decision-making isn't just about saving time and 'getting through' all the decisions fast enough; it's sometimes about not having to give a tough job to a human, about not putting the pressure on a human to make a decision that might end up being the wrong one. It's about not being capable of – or simply not wanting to – deal with the guilt, the backlash, the social issues around making incorrect decisions that harm someone else. It's about decision paralysis, not wanting to be burdened with the repercussions after the decision is made. It's about convenience, status, simplicity, lack of care; 'anything for an easy life', if you will. It's not just about speed and productivity.

A lot of the time, we know what happens as a result of our decisions. We know that buying more items with plastic in them results in more plastic in the oceans, which results in more dead fish – but we buy plastic anyway, because decision-making is not always so simple. We can't optimise for absolutely everything (in this case: the environment, our time, our money, convenience, and so on). Sometimes, the use of AI is about relieving ourselves of the burden of having to analyse the world and make a decision (and live with the consequences) as a result. But that yearning for simplicity can so easily lead to bad decisions being made, and AI systems provide an 'effective outlet' for that blame, as opposed to the people who made an active decision to use it. Instead of working out what we as individuals think is the best thing to do in difficult cases, we are looking to computers to do it for us.

'Decision-making on tap' can be hugely powerful, especially if those AI-powered decisions help save lives by spotting cancer earlier on scans, or help reduce our energy usage by predicting and managing the supply and demand of our national grids. If we're moving towards a world where we outsource our decision-making to computer systems built by previous generations, and never really consider what actions we're taking in real time, with that must come huge responsibility in ensuring what we build is not half-hearted.

Will AI 'Take Over'?

One common question asked about AI when it comes to decision-making is: 'At what point will these systems "take over"?'

You might have heard of the singularity. It's the theoretical point at which AI systems become so intelligent they surpass human intelligence, meaning they can then start building and improving themselves without the need for our input. It's normally the inciting incident in any dystopian sci-fi story centred on AI. Some researchers and companies are focusing on what's called Artificial General Intelligence – building AI systems to be able to learn anything (hence, 'general'), as opposed to one single task (this is often called Artificial Narrow Intelligence) – with the mission being to have more powerful AI systems, which can do more, better, work for us without needing as much human input. There's a fear that if and when we manage to build AGI systems, they will naturally lead to the 'super' systems of the singularity.

There is much debate in the field and in the popular media about if and when AI systems will overtake humans intellectually. And not to dismiss it entirely as a worry (as it rightly is), worrying about 'robots killing us as a result of the singularity' is utterly pointless if we don't also worry about, and hold accountable, those building the systems of today. What's built now leads to what's next. Using the *Terminator* reference is dangerous in two ways. It's either diminishing to the great, useful technology being built, which has the potential truly to revolutionise society for the better (and using the term turns people against something they should be backing), or it's diminishing the size of the problem and the impact it may have on the world ('we know it's bad so let's make one overly simplified joke about it and then move on with our lives without actually doing anything about it').

The singularity is concerning only if we don't hold accountable those playing an active role in building for it.

WHAT IS RIGHT AND WRONG?

AI is an accentuation of humans. If we truly are building machines to do our heavy-thinking as well as our heavy-lifting for us; if machines are going to be better than humans in making (well, advising on) complex moral decisions, we need to work out how to make the dataset, the algorithmic recipe and the initial instructions that the humans dictate align with what makes most sense for society.

This amounts to the question, to which each of us has a different answer: 'What is right or wrong?'

Right now, the rules of right and wrong are being decided by the AI builders. The moral rules that are being coded into the AI systems will power future decision-making. How the AI systems prioritise things is also being decided right now. Take plastics for example. If you have to make a decision about whether to buy a plastic bottle, you have to weigh it up in terms of your own preferences: do you put the environment, or your own convenience first? Decisions are being made – by humans – about what is most important to optimise for, and AI systems will learn that this is the North Star if it's at the core of its build.

For example, IBM is working on a piece of research called Project Debater.[18] It is building an AI system that will be able to debate humans on complex topics – the idea being that it will help people make complex decisions by 'having it out' with them; facilitating 'better' debate, combating misinformation and providing more well-informed arguments. Sounds pretty useful, and I don't deny that kind of work will do some real good in the world. But here's a question: can you form an argument without having a concept of what is right and wrong? And if not (which

I suspect), how do you get a computer to know right from wrong? And whose version of right and wrong?

Take, for instance, a debating topic such as this: 'Is religion good for the world?' Now, that's a multifaceted question and a broad ill-defined topic open to many answers. We could boil it down to working out how impactful a strong religious community is on the world versus how devastating religious conflict is. A computer would have to know how to weigh all the variables that go into that – the value of charity, the value of human interaction, the value of productivity (and in what terms – monetary versus societal survival, for example), the value of human life. A computer system can't come up with that itself – and I'm willing to bet that reading all the literature ever published and every word posted on the internet wouldn't give a system enough of an answer (which is essentially what IBM AI systems are able to do). Coding 'lived experience' into the systems means coding in the thoughts, emotions and morals of the person or people building it – and these types of questions cannot be answered with logic alone. The system would have to have a concept of what right and wrong is, in order to sift through the information and weight it according to 'good' and 'bad'.

There's a branch of social psychology called 'moral foundations theory'. It attempts to explain how and why human morals differ, based on how sensitive we are to each of the six moral foundations.[19] There's 'care', opposite to 'harm'; 'fairness', opposite to 'cheating'; 'loyalty', opposite to 'betrayal'; 'authority', opposite to 'subversion'; 'sanctity', opposite to 'degradation'; and 'liberty', opposite to 'oppression'. Research found that political leaning can be tied to which of the foundations each person is more sensitive to – conservatives emphasise loyalty, authority and sanctity; liberals emphasise care, fairness and liberty.[20] The idea of 'good' therefore can be broken down into what you as an individual consider most important.

When it comes to 'coding in ethics' to an AI system, there have been proposals for 'tagging' various actions and events with their relevant portions of the six moral foundations and giving that as a base database for the AI system to build upon and learn from.[21] Something like 'telling your friend they look nice when you don't actually think it, to make them feel good', could have a high score in care and loyalty, and lower score on fairness, for example. If you could build a database with enough actions that have been defined and checked by enough diverse humans (so as to account for differing opinions), the theory goes that an AI system could then start to learn what's broadly agreed to be right and wrong by all – and then piece together all these bits of information when presented with one big meaty question to answer. It means that it could, at best, result in human-level moral decision-making, or it could mean that by aggregating many humans together – as opposed to just one human – you'd get a 'more moral' system and therefore one that's better equipped to be making moral decisions than any person alone.[22]

Either way, we're again left with that initial dataset, that core information on which the system is built, and we don't always get to see what's under the hood. IBM, or anyone else building any kind of AI system that may have to make moral decisions (i.e. most of them), might not ever disclose its root data, and so we wouldn't know if the tagging was sufficiently diverse or broad. We wouldn't know if it even used such a method of coding in ethics in the first place.

The key thing to pay attention to is not the answer to 'Is religion good for the world?', but rather, 'Based on what?'

AI systems are a continuation of our thoughts, our practices, our ethics – it's a reflection of us, a summation of us as a society, of the people who built it. It's also therefore our responsibility – it's something we create; it's not a separate autonomous being with regard to thought.

The cost of us not seeing that it's an extension of us, and of not realising that handing over decision-making to AI systems doesn't mean we also hand over responsibility, is a future where control is lost, the world is defined by the few, and we have less say in the here and now. That may sound like utopia for those in control now, but for those who are not, AI systems present a problem in accentuating the existing unfairness in our human society.

Many people are building AI systems right now, and the only way for them to be held accountable is for more people to ask the questions, add the insight and diversify the data that those deep in the build may be missing, whether deliberately or not. We are in control of AI – and letting the idea that we're not continue to blinker us might, ironically, be the path to the self-fulfilling prophecy we fear.

Beyond the Aliens
in Astrobiology

In 1996, an astonishing paper was published in the journal *Science* claiming to have proven that there was life on Mars.[1] NASA biochemist David McKay led the team behind the findings, which were based on the analysis of a fragment of an ancient Martian meteorite called ALH84001, found in the Alan Hills of Antarctica in 1984. The team's scans of the meteorite had unearthed so-called 'biogenic features': markings, channels and structures throughout the rock, which they interpreted to be caused by bacteria-like lifeforms.

Meteorites are rocks from outer space that have travelled through the Earth's atmosphere and landed on our planet. They could be pieces of shattered asteroids, rocky cores of comets, or fragments from the surface of planets or the moon, which break off when large asteroids or comets collide with them.

The beauty of ALH84001, other than it being only one of the few Martian meteorites that were known to scientists at the time, was that it was much older than the others. At 4.5 billion years old, it came from a time in history when Mars likely had liquid water on the surface: the period during which life was far more feasible on the red planet. The announcement was arguably the most exciting one from NASA since the Apollo missions. Carl Sagan remarked: 'If the results are verified, it is a turning point in human history.'[2]

The paper not only piqued the interest of those in the field, but the then president of the United States Bill Clinton made a televised announcement the day after the paper was released, on the South Lawn of the White House, where he told the world of the possible discovery of life on Mars.[3]

The paper, as you might imagine, captured headlines all over the world; the *New York Times* ended its piece 'Life on Mars?' with the line: 'given the intergalactic speed with which science has advanced in this century, it would be prudent to hold the jokes'.[4] The legitimatisation of searching for extraterrestrial life was a story in and of itself.

It wasn't long, though, before the scientific consensus became more sceptical. McKay's team had put forward four arguments for why they believed that ancient microbes were fossilised and preserved in this rock in four specific features they believed had to have been created by living things. But for each argument, researchers were able to give examples where non-biological processes had created the same specific features.

One particular line of argument was a tougher nut to crack, as it were, which was the existence of magnetite mineral crystals in ALH84001. Some ancient bacteria found on Earth produce these magnetite crystals, which indeed did look very similar to those found in the meteorite, and so McKay posited that maybe Martian bacteria did the same. For years, this was the reasoning McKay touted as the evidence of ancient life on Mars, as no one had been able to produce anything similar artificially.

Until 2004, when his own brother Gordon McKay, who worked down the hall at NASA, did exactly that.[5] Gordon McKay and the team of scientists he worked with not only managed to create a batch of similar magnetite grains to those found in the meteorite, they did it in a lab where they simulated the conditions that the meteorite is known to have experienced on Mars all those years ago. David McKay argued that the crystals weren't similar enough to those his team found in ALH84001, and indeed he continued to insist that his findings strongly suggested life on Mars, but the rest of the scientific community wasn't convinced.

David McKay was a respected scientist, though, and the publication of his results and the subsequent questioning of them by the broader community are a demonstration of science working at its best. Clinton, in his enthusiastic backing of the American space programme's search for evidence of life on Mars, represented a shift in support, funding and credibility of the emerging field of astrobiology. Despite the scientific consensus that ALH84001's markings didn't correspond to ancient Martian life, the mainstream hype and presidential treatment opened up discussion about the findings even further and gave them the respect the science of 'life elsewhere' had rarely received in the past, both from the public and scientists across fields.

In the process of picking apart the findings, and showcasing non-biological explanations for each piece of evidence put forward in the paper, researchers developed standards that would allow them to evaluate the presence of life in other Martian samples, and build the basis of questions, approaches and methods of proof for the astrobiologists who followed McKay and his team.

In some way, this seems like the way astrobiology, and the scientific search for life elsewhere, should always play out in terms of both the public and academic reaction to new hypotheses presented in the space. Excitement, healthy scepticism, respectful consideration and interrogation of the research presented, and increased understanding across the board regardless of whether the ideas are confirmed or not, make for stronger science.

Unfortunately for astrobiology, despite its increased credibility since the Clinton speech, there is still one word that plagues both its mainstream perception and our openness to the broader implications of the field: 'aliens'.

ALIENS RUIN SCIENCE

It's no secret that sci-fi writers love aliens. The idea of extraterrestrial beings coming to Earth to take over our planet, or existing in many forms across the universe, and who we can visit, fight and fall in love with, makes for incredible world-building, intriguing thought experiments and plenty of exploration into the way humans think, act and find their sense of meaning. It's no wonder then, with such rich storytelling and much self-questioning during the process of reading or watching these tales, that some people link their perception of what aliens might be like to explanations of strange things they experience in everyday life.

These stories make news. We love hearing about how people try to explain their extraordinary experiences; whether that's in a cruel, mocking manner or simply out of natural curiosity. People who claim to have seen UFOs, or claim that the government is hiding knowledge about extraterrestrial life, or that their fields turned into some kind of alien street art, often have their views disregarded by the mainstream. Without proof deemed strong enough for public consensus, such as videos or photographs or material samples, these ideas are often immediately put in the 'not real' pot. The problem, though, is that they are all lumped into that pot without engagement in the rationale, specificity and data behind them. This means that astrobiologists need to be overly cautious in their research announcements, for fear that their work will be reduced to the headline: 'Aliens Might Exist'.

The thing is, though, aliens *might* exist. We don't know. And unfortunately, the scientific quest to work out the answer can be plagued by the claims made by people without evidence, making it a somewhat fragile field of study. There's a difference between making a claim and then backing it up with some form of rationale in order to be believed, and saying something without evidence and assuming others will take it at face value. Christopher Hitchens wrote in his 2007 book *God Is Not Great* a phrase that would go on to be known as 'Hitchens's razor': 'What can be asserted without evidence can be dismissed without evidence.'[6]

Good scientists *do* present evidence; but if we dismiss the ideas before even considering their rationale, simply because they concern something we believe to be 'not real', we risk ignoring those who are dedicating real effort to answering possibly the most profound question we as humans have: 'Are we alone?'

THE SCIENCE AND PSEUDOSCIENCE OF PANSPERMIA

What makes astrobiology even tougher to be understood and pondered by members of the general public is that some very credible scientists say some very non-credible things about extraterrestrial life, and refuse to update their own knowledge when proven wrong, meaning sometimes it can be hard to know who to trust. This fragility is exemplified in the discussions around a particular idea in astrobiology: panspermia.

Panspermia is the hypothesis proposing that bacteria and other life forms can be transported through space to Earth on bodies such as comets and meteoroids, and therefore life exists throughout the universe and could have been 'seeded' on Earth. The scientific side of the theory prompts questions such as 'What kind of life forms could survive the journey through space and the rough entry into Earth's atmosphere?' or 'What kind of more favourable conditions might have existed on another planet that resulted in the formation of life in the first place?' These questions tie in with fascinating research areas such as extremophiles, organisms that thrive in physically or geochemically extreme conditions detrimental to most terrestrial life, such as in extremely high or low temperatures, alkaline or acidic systems, or extremely dry areas. Some scientists are working out the mechanics of the rock ejection from other planets, the journey through space and the impact with the Earth's surface, working out at which angle the impact would have to happen to avoid destruction of life onboard and so on. In short, the panspermia hypothesis is an intriguing one that can be explored scientifically and, in that process, increase our understanding of life in the universe as a whole.

Where the problems with panspermia come is when the concept of 'directed panspermia' is brought up. This is the idea that instead of a rock, by chance, travelling through space with life on board and impacting the Earth, seeding life as we know it in the process, another intelligent lifeform somewhere else in the universe *deliberately* planted the seeds of life here on Earth. This idea has garnered much attention as it was proposed in the 1970s by Nobel Prize-winning scientist Francis Crick (best known for co-authoring the paper that proposed the double helix structure of DNA with James Watson), along with a chemist Leslie Orgel.[7] In 1981, Crick wrote a book about his theory.[8] The idea is widely considered speculative and unscientific as the arguments put forward don't seem to be grounded in evidence that can be proven or easily disproven.

Nowadays, there are scientists floating an idea called 'cosmic ancestry', based on the panspermia views of Chandra Wickramasinghe and Fred Hoyle, which maintains that life has no date of origin, and so has always existed, and can only descend from ancestors at least as highly evolved as itself. The idea that life has always existed is contrary to almost all contemporary scientific views, and without any real evidence being put forward along with the claim, it is regarded as pseudoscience. Again, these were two scientists who were well respected in other areas of their work, proposing ideas in such a way that the wider scientific community either ignored or rejected their views. Some of their papers have been debunked in peer review but they continued to stand by their unfounded claims, creating their own journal and institute, where they could publish their ideas to their own audience. Wickramasinghe, the now deceased Hoyle and another panspermia proponent, Milton Wainwright, have collectively made many claims that have been debunked and criticised for their pseudoscientific approach. They have claimed that viruses and bacteria enter the Earth's atmosphere from space and have been responsible for major epidemics throughout history, an idea that is widely rejected by the scientific community.[9] They have claimed that the fossil of the oldest known bird, *Archaeopteryx*, was a forgery, which palaeontologists called 'absurd' and 'ignorant'.[10] They claimed that the micro-organisms that Wickramasinghe's research team found in the stratosphere came from space, despite there being no evidence that the micro-organisms came from space and not from Earth.[11] Their 2013 research on those micro-organisms was published in the *Journal of Cosmology*, a journal criticised on several occasions for publishing dubious science, and was covered with much fanfare by the *Daily Mail*, the *Telegraph* and the *Independent*, proclaiming the discovery of alien life.[12]

When newspapers cover astrobiology stories, understandably they use bombastic headlines and are more interested in scientific papers that make big claims. But covering pseudoscience without checking in with the broader scientific community and then essentially publishing falsehoods is not only incorrect and entirely misleading, it puts the public off believing any further stories from the field when they discover these initial stories are bogus. It's confusing to say the least, and it's no wonder astrobiologists can be critical of how the media cover their work.

There's a secondary issue with this beyond the wrong information going out: the public, through their hesitation to believe in bad science, are essentially being left out of the fascinating scientific discussions that result from the panspermia hypothesis. For instance, if life on Earth did indeed originally come from a meteor laced with Martian life, then how did life originate on Mars? And how did that life manage to survive the treacherous journey? And could Earth-based bacteria have travelled on an ancient ejected rock onwards to elsewhere in the solar system or beyond? Or could the rovers and other payloads we've landed on the moon, Mars and Venus have had some kind of terrestrial lifeform on the outside of their metal casings, with which we've now contaminated other places?

Panspermia is possible, but we don't really have a way of proving whether it's probable, and when we're talking about these ideas we must consider the difference between the two and not just jump to conclusions, as well as try hard to work out whether the science is credible.

How to Tell Science from Pseudoscience

Surely, though, we can tell the difference between 'crazy' ideas that have no real form of proof or concrete reasoning behind them, such as directed panspermia, and those that, like the ALH84001 claims, are proposed with reasonable rationale, which can then be tested and confirmed or rejected. Surely, when we can separate those two very different kinds of ideas, or at least, very different kinds of presentation of those ideas, we cannot be fooled or tainted by that which adds nothing to the useful body of knowledge.

Of course, there are many counter-examples to this idea that we can separate reasonable ideas from those without backing, such as the community of people who believe in 'alternative medicines' and the so-called 'Flat Earthers'. Beyond the ideas known to the masses as being pseudoscience, though, when faced with something new, sorting science from pseudoscience can be a difficult task for reasonable people.

New ways of thinking about the world are absolutely welcomed in the scientific process, if they can be justified in such a way that they can ultimately be proven; this characteristic, arguably the thing that makes science 'science', is called 'falsifiability'.

Falsifiability as a concept was introduced by philosopher of science Karl Popper in the 1930s, and it essentially means that any scientific hypothesis must be inherently testable, and one that always remains tentative and open to the possibility of being wrong.[13] Claims need to be able to be tested using tools that already exist or we can imagine building, or they need to be able to be observed somehow repeatedly. Over time, the claims get fine-tuned, or elaborated on, or discounted, and knowledge increases as we continue to scrutinise and check. There's an element of this process that is quite impractical, as the burden of proof is high and we need to be able to build on top of ideas and not get too caught up in continually proving them, but ultimately if we can't even approach a hypothesis knowing how to prove or disprove it, then it is not within the scope of science.

An example of a falsifiable claim would be 'all chairs are painted red', as you would only need to observe one chair that is not red in order to disprove the theory. An example of a non-falsifiable claim would be 'my red chair can speak to me', because if it is wrong, you would only ever be able to find an absence of evidence: no words said while you're in the room, no recordings of the voice from the original claimant, and so on. In order to disprove it, you'd need to do an exhaustive search of all the possible situations in which a chair might speak, which is arguably an infinite task. These kind of statements simply aren't scientific.

Checking whether a statement is falsifiable isn't enough, though, to work out whether something is good science or not. Massimo Pigliucci, professor of philosophy at City University of New York, wrote about the limits of falsifiability in his 2010 book *Nonsense on Stilts: How to Tell Science from Bunk*.[14] In short, he argues that Popper's idea is too simplified for the messiness that the real world exhibits and, when taken to the letter, would disregard good science. For example, Pigliucci uses the example 'all dogs have four legs', which is not true for dogs born with genetic defects, but that information obviously shouldn't discount the original idea.

Pigliucci suggests instead that we need to find other ways of distinguishing, as he puts it, 'science from bunk', if we're to work out what is true and untrue without becoming deep experts in every topic we wish to ponder. For starters, he defines what makes science different is naturalism, theory and empiricism. He writes: 'What all scientific inquiry has in common, however, are the fundamental aspects of being an investigation of nature, based on the construction of empirically verifiable theories and hypotheses.' So anything that is supernatural, and has no underlying theory or hypothesis as to why it is what it is, and cannot be tested, cannot be considered science.

Pigliucci also writes about how we defer to experts in order to know if ideas are legitimate or not, and so we must be able to tell if someone is indeed a reliable expert. He refers to philosopher Alvin Goldman's five-point guide to figuring out whether to trust a prospective expert or not: (1) an examination of the arguments presented by the expert and his or her rival(s); (2) evidence of agreement by other experts; (3) some independent evidence that the expert is, indeed, an expert; (4) an investigation into what biases the expert may have concerning the question at hand; and (5) the track record of the expert.[15]

Another way of assessing what is a non-scientific approach versus credible science is how open the claimant is to accepting new evidence contrary to their own. Caltech theoretical physicist Sean Carroll, in his book *The Big Picture: On the Origins of Life, Meaning and the Universe Itself*, writes about the importance of updating our knowledge in our quest to find truth. Carroll uses Bayes' theorem, a formula that describes how to update the probabilities of hypotheses when given new evidence, which has become a popular formula in philosophy of science to determine good science from bad. As Carroll puts it, Bayes' theorem is a way to think about credences, our belief in or acceptance that something is true, and how new information about the world changes the credences we have assigned to different ideas.

What's key about considering Bayes' theorem is that it gives us hints as to how belief works, and why sometimes it cannot be altered. Prior beliefs on the part of the claimant, if they are very strong, are unlikely to change regardless of the weight of new evidence put to them. Carroll puts it nicely when he writes, 'Everyone's entitled to their own priors, but not to their own likelihoods': we can originally believe whatever we like, but it's our responsibility to be open-minded and humble enough to accept new evidence when it tips the likelihood of truth out of our favour.[16] He also speaks about all evidence mattering. For example, we cannot only take into account an eyewitness account of spotting a UFO as the key evidence, we must also consider all of the searches for UFOs that have come up empty, and the fact that many photographs put forward as evidence have been found to be hoaxes, and so on. All evidence needs to be brought together; we can't pick and choose the evidence we want to consider.

Working out science from pseudoscience, then, is a hard task. What makes it even tougher, though, for those working in astrobiology is the difficulty in taming the media and public sentiment from immediately deferring to the terminology of aliens and UFO pseudoscience when describing credible scientific work in finding life beyond our planet. The two areas get inextricably linked and are then difficult to disentangle, and this doesn't give the non-expert member of the public much confidence in working out what is believable. It can make the field something people don't want to admit they are interested in, for fear of coming across as someone who has fallen for pseudoscience.

WHAT DO ASTROBIOLOGISTS DO?

The field of astrobiology is concerned with life in the universe, which can more specifically be bundled into three key questions: How does life begin and evolve? Does life exist elsewhere in the universe? What is the future of life on Earth and beyond?[17]

Recently, the focus has been on 'habitability', in other words, what makes the Earth habitable; how, when and why did it become habitable; and are or were any other bodies in our solar system habitable beyond the Earth? More broadly, scientists are investigating the origin and evolution of planetary systems, and seeking out biosignatures to detect life or evidence of a history of it – clues such as atmospheric gases or organic matter formed from biological processes, or minerals and fossils found which indicate life (similar to what McKay was trying to find in

ALH84001). They might look for these biosignatures on bodies in our solar system, such as Mars, Jupiter's moon Europa or Saturn's moons Enceladus and Titan, or by using spectral imaging of planets outside our solar system, called exoplanets. Astrobiologists might also look at the 'steps to life', in other words, tracking how exactly life evolved on Earth as well as how the physical environment evolved around life; or maybe finding non-biological sources of organic compounds to work out how life might have started in the first place.[18]

Astrobiologists focus their time on finding microbial life, as opposed to signals or evidence from intelligent species capable of communicating with us. This might feel less 'sexy' than seriously searching the skies for aliens (we'll get to this), but the repercussions of having more knowledge about life in our universe go beyond simple curiosity. For example, understanding how our environment changes with life can give us clues about future climate change; and building instruments, such as sensors and landers and imaging systems, to measure extreme environments leads to new technologies that can be used for answering other difficult questions here on Earth.

When NASA Cries 'Alien'

Despite the broader, near-term, real-life implications of the field, it seems that it's a tie to extraterrestrial life that makes the biggest news, and this is what NASA tends to put more effort into communicating. Knowing the public are more likely to get excited about the answer to the question 'Are we alone?' being a possible 'No', backed up by the latest findings, can also easily backfire when there's a disconnect between the public understanding of what goes on in astrobiology and what they see and hear in popular media and entertainment.

A common response to NASA announcements declaring some form of evidence for extraterrestrial life on the red planet, usually based on the discovery of water, is: 'Haven't we already found water on Mars?' And it's understandable why the general public might not be able to follow the multiple announcements when every one leads with the same headline and sentiment: we've found evidence of water and therefore possibly life on Mars. A quick Google search presents headlines along the lines of 'NASA Confirms Water on Mars' from 2012, 2013, 2015, 2018 and 2019.[19] Without reading each in detail and following the field over time, it's not clear at all how important each announcement actually is.

Without the detail, such as from what time period, or in what section of the planet and what kind of water body, and therefore the corresponding chances of current life versus ancient life, it's easy to lump all the announcements together into one idea, which quickly starts to feel old.

NASA also has a tendency for making announcements in widely distributed press releases about . . . an upcoming press release. This pre-press release press release does wonders in terms of getting people excited about what's to come: something that requires an announcement about its announcement must be huge news, right? But often the public are left disappointed when the announcement doesn't seem all that remarkable. Without the understanding of exactly where we're at in the astrobiology field, and the types of questions scientists are focusing on, NASA and the scientists making these announcements fail to manage the expectations of the public and media; leading to further disappointment, misunderstanding and confusion about what on Earth astrobiologists are actually doing.

Of course, NASA needs to keep the excitement for space as alive as can be. It's a publicly funded endeavour after all. But those on the inside can sometimes fail to step outside of their understanding of the field, and their wonder for science fuelled by a deep knowledge of it, and miss the fact that most people think of this science either as pseudoscience, old, going nowhere or simply not important.

There's a worry then that in the pursuit of making astrobiology exciting for the public so it can continue to justify their funding for future missions, NASA risks crying wolf by overpromising on what missions are built for or what research findings have uncovered, and losing the general support from the public in its entirety.

Balancing hype, overhype and alignment with pseudoscience seems to be a tough task when it comes to sharing astrobiology with the world, but considering the fact that this field asks possibly the most fundamental questions humans have, 'Why are we here?' and 'Are we alone?', it seems ludicrous that there's even any need for discussion on how to better communicate the field to the public.

Some might argue that, once you understand really what's going on in the astrobiology world, it becomes less interesting. The reality of how little we know and how few our successes are in answering these questions; the deliberate focus on microbial life as opposed to intelligent life; and the many caveats that come with each new piece of research, all might result in a disinterested public.

Questions such as what are the chances of finding life elsewhere, how do we actually go about finding it, what exactly are we looking for, and what does it mean if and when we find signs of it, are inherently interesting. Deferring to the 'alien' narratives in pursuit of excitement both cheapens and confuses the message, to the detriment of everyone's understanding and the support of some of the most crucial science being done.

THE SEARCH FOR EXTRATERRESTRIAL INTELLIGENT LIFE

Using the word 'alien' when talking about the science of finding life elsewhere does lead to disbelief and the weakening of the message, but the irony is that this is exactly what the scientists are trying to do: find alien life of some form.

So maybe the issue is not the use of incorrect language or the easy association with science fiction in order to drum up public excitement, but about how we, as those on the outside, perceive those who are genuinely, scientifically, trying to answer the question about our place in the universe.

In 1961, astrophysicist Frank Drake penned an equation to stimulate discussion at the first scientific meeting on the 'search for extraterrestrial intelligence', the field as a whole now known as SETI. The equation was to act as a 'roadmap' of sorts for what scientists would need to learn in order to find the answer, which essentially defines the number of intelligent civilisations in our galaxy, N, as the product of various astronomical, biological and cultural factors:[20]

$$N = R_* \cdot f_p \cdot n_e \cdot f_l \cdot f_i \cdot f_c \cdot L$$

Where:

R_* is the average rate of star formation in the galaxy.

f_p is the fraction of those stars that have planets.

n_e is the average number of planets that can potentially support life per star that has planets.

f_l is the fraction of planets that could support life that *actually* develop life at some point.

f_i is the fraction of planets with life that go on to develop *intelligent* life.

f_c is the fraction of civilisations that develop a technology that releases detectable signs of their existence into space.

And L is the length of time for which such civilisations release detectable signals into space, in other words, the lifetime of a civilisation.

There are various estimates about what each factor might equal, thus giving different estimates for N ranging from 20 to 50 million.[21] The last four terms are unknown, and are the focus of many an astrobiologist's work, but this means that working out statistical estimates is impossible. Whether you take an optimistic or pessimistic approach to your estimates, there seems to be a clear rationale for at least trying to find these extraterrestrial intelligent civilisations. The equation seems to point to their existence; 'our universe is surely big enough for us not to be alone'.

Then there's the famous Fermi paradox. Named after physicist Enrico Fermi, it's the contradiction between the high-probability estimates in more optimistic versions of the Drake equation, and the lack of evidence of existence of extraterrestrial beings, or more succinctly put: 'Where are they?'

Whichever way you consider it, the SETI field is speculative, which makes it different from what many would consider science. Where astrobiologists concern themselves with hypotheses that fit firmly into existing scientific theories, SETI requires a guess of sorts that intelligent life is indeed out there. Even with credible scientists working on radio tele-scopes to seek out signs of intelligent life, and using credible methods paired with scientific scepticism, the speculative nature of SETI means it sits in a grey area between the definitions of science and pseudoscience. As the editors of *Nature* wrote on the fiftieth anniversary of SETI efforts: 'SETI is arguably not a falsifiable experiment. Regardless of how exhaus-tively the Galaxy is searched, the null result of radio silence doesn't rule out the existence of alien civilisations.'[22]

'LITTLE GREEN MEN'

The SETI field is no stranger to the so-called 'giggle factor': it is plagued by it.[23] In 1993, the US government's NASA budget removed SETI efforts from its federal funding, after which Nevada Senator Richard Bryan famously boasted: 'The Great Martian Chase may finally come to an end . . . As of today, millions have been spent and we have yet to bag a single little green fellow. Not a single Martian has said 'take me to your leader', and not a single flying saucer has applied for FAA approval.'[24] He wasn't the first member of Congress to mock and ultimately curb fund-ing for SETI efforts, following in the footsteps of Wisconsin Senator William Proxmire who, in 1982, amended NASA's budget to curb early SETI efforts and handed the space agency one of his 'Golden Fleece' awards – an annual callout he did for federally funded projects he person-ally believed were useless.

With SETI sitting in this grey area of speculative science, it is possible to understand why Congress doesn't want to fund efforts and run the risk of looking bad in the eyes of the public, but when you consider the fact that other research areas such as gravitational waves, string theory, dark matter, the Higgs boson and black holes deal with sometimes spec-ulative phenomena and have taken years of research to gather compel-ling evidence, it's clear that it's the alien link to SETI that is sometimes too much for those not in the know to handle. As astronomer and astro-physicist Professor Jason Wright and anthropologist Michael Oman-Reagan astutely wrote in the *International Journal of Astrobiology*, big sci-fi-esque claims don't seem to plague other areas quite so much: 'Indeed, discussion of the only (apparently) seriously funded effort for a permanent human presence beyond Earth, the SpaceX aspirations for a Mars settlement, is pervaded by fancy, including talk of planetary apocalypse, tyrannical runaway artificial intelligences and supervillains.'[25]

Instead, SETI efforts are privately and philanthropically funded, with the SETI Institute attracting backers such as Microsoft co-founder Paul Allen, who donated $30 million to build the Allen Telescope Array in 2007, and Russian billionaire Yuri Milner pledging $100 million in 2015 for SETI projects over the following ten years.[26] Most of the efforts focus on using radio telescopes to scan the skies for transmissions from some form of intelligent civilisation. The signals they are looking for could, of course, be signals sent deliberately from another civilisation, or simply by-products of their own communication systems, much like our own TV and radio transmissions on Earth.

And SETI efforts aren't all that expensive, especially when you consider the potential upside of finding a positive result of some kind. Surely it's worth having a few radio telescopes around the world listening for signs of intelligent life, manned by people who will analyse the results with scepticism, verification, peer review and scientific protocols . . . just in case?

THE FASCINATING QUESTIONS WE DON'T KNOW TO PONDER

Even within the scientific community, there are those who sneer at the idea of extraterrestrial life research. Some scientists stay away from any kind of association with SETI for fear that they'll be branded a crank, and have their scientific career shortened as a result. For others, SETI is simply tolerated; seen as some kind of nifty side project running alongside the 'proper science'.

On the other hand, some scientists will loudly discredit the exploratory work of SETI scientists, insinuating that they are verging on pseudoscience, which can not only destroy the reputation of credible scientists but further confuse those on the outside as to what is reliable research.

The result of this is that it effectively isolates the fascinating, fundamental questions that come with accepting that the search for life elsewhere is fair game – both from a SETI and an astrobiology perspective – from those who have been alienated by the alien narrative.

There are the more scientific questions, such as how exactly do we define the life we're searching for? If we look for biosignatures that make sense for carbon-based beings like ourselves, do we miss out on searching for life based on silicon, for example? How do we design experiments to search for this alternative form of life? And if we find this alternative life, what does that then mean for our understanding of how life starts and evolves?

There are the questions around how SETI researchers define intelligent life signals, in terms of what makes them signals that are signs of life versus some other cosmic event. And how do we choose the most promising search areas? And how do we actually know if we're right when we do find something resembling what we believe to be an intelligent life signal?

The questions go beyond scientific or technical questions into the realms of the social sciences, with astrobioethics researchers considering the moral implications of taking humans to Mars, or how cultures and religions might change if the question 'Are we alone?' was answered one way or the other.[27]

TALKING TO ALIENS

One particularly fascinating question, which extends far further than science and into the realms of both social sciences as well as individual opinion, is simply, 'Who are we'?

On 16 November 1974, in the tropical forests of Puerto Rico, astronomers, government officials and a few other dignitaries gathered to witness what would become the most famous deliberate transmission of a human message out into space. Dreamed up by none other than Frank Drake himself, with the help of Carl Sagan, the message sent from the radio telescope at the Arecibo Observatory, known as the 'Arecibo message', started its 25,000-year journey to a cluster of stars called M13, in the hope that something, or someone, somewhere might eventually take our call.

It was a 168-second broadcast, made up of binary digits representing some basic information about humanity and Earth. It included the numbers one to ten; the atomic numbers of hydrogen, carbon, nitrogen, oxygen and phosphorus (the elements that make up DNA); a representation of the helix structure of DNA; a picture of a human, the dimensions of an average man and the human population of Earth; a picture of the solar system, indicating Earth as the source of the message; and even a picture of the Arecibo telescope they used to transmit it.

The Arecibo message was less an attempt at real conversation with extraterrestrial beings and more a demonstration of the technology at the time, but it prompted a global discussion about the merits and pitfalls of flagging our existence to anyone else who might be beyond our planet. The Royal Astronomer of the United Kingdom, Martin Ryle, wrote that we'd be inviting catastrophe from 'any creatures out there [who might be] malevolent or hungry'.[28]

The conversation would pick up every now and again whenever efforts to 'make contact' arose, such as NASA's 1977 Voyager probe, which left with a 'golden record' on board with songs and messages from Earth, or the agency's 2008 transmission of the Beatles song 'Across the Universe' using a 70-metre dish in Madrid, Spain. These efforts were broadly considered unlikely to work, though, so the debate was never quite seen as a matter of international safety.

In 2015, the discussion entered the realms of serious debate when Douglas Vakoch, an astrobiologist who had previously spent sixteen years at the SETI Institute, spoke at the American Association for the Advancement of Science (AAAS) about the need for 'active SETI' as opposed to just passive listening. A few months later, METI (Messaging Extraterrestrial Intelligence) International was set up, led by Vakoch, with a mission to explore the design and transmission of interstellar messages.[29]

Vakoch's message didn't go unchallenged, with a small group of prominent academics and space experts, even including Elon Musk, penning a petition urging that the decision whether to transmit 'must be based upon a worldwide consensus, and not a decision based upon the wishes of a few individuals with access to powerful communications equipment'. They argued: 'As a newly emerging technological species, it is prudent to listen before we shout.'[30] (It's worth pointing out that some of the signatories of the petition were advisors to other messaging projects, so competition may have been at play.[31])

The argument that messaging extraterrestrial beings represents an existential threat in that it might attract the attention of potentially hostile alien civilisations can be put alongside the argument that we've already been broadcasting signals since the dawn of radio in the 1920s. But the METI researchers admittedly are arguing for more deliberate messages and stronger signals focusing on the areas of the universe with more likelihood of intelligent life. The proponents still champion the cause regardless. Seth Shostak, who leads the search for extraterrestrial life at the SETI Institute, wrote in the *New York Times*: 'I, for one, would hesitate to let a paranoia based on nothing more than conjecture shackle the activities of our children and our children's children. The universe beckons, and we can do better than to declare that future generations should endlessly tremble at the sight of the stars.'[32]

How Do We Communicate Earth?

For me, the most intriguing question surrounding the METI debate is not whether we'll be obliterated by aliens or which team should get the funding, but rather, what should we say?

As Vakoch said to *Forbes*: 'It's too late to conceal ourselves in the universe, so we should decide how we want to represent ourselves', and it's this discussion that transcends the world of science into that of society and the self.[33]

If you were to pen a message to an extraterrestrial intelligent civilisation, what would you say?

Would you tell them about our wars, our genocides and our injustice, or would you only tell them about our triumphs, our progress and our dreams?

Would you transmit the entire internet, as Dr Shostak has suggested, to let the extraterrestrial civilisation work us out for themselves, or would you rather we curate the information sent?[34] Can we encapsulate the billions of people on Earth in just one message? And if we curate that message, who does the curating? Who speaks for Earth? The politicians, the scientists, those who just get out and do it without asking anyone? You?

And how would you translate that message so it makes sense for a species that doesn't speak languages in the same way humans do? Morse code? Binary? Symbols?

These are not questions that can be answered by scientists alone, and they are also not questions to sneer at. We already have organisations on Earth transmitting these messages; don't you want a say in what is being said about us?

WORKING OUT WHO WE ARE

In order to work out what to transmit, surely we also have to work out who we are. And engaging in the METI debate isn't the only way to end up questioning the very nature of our existence within astrobiology and the search for life elsewhere.

When you consider the question 'Are we alone?', there are only two finite answers.

If we find out that we are alone, what does that mean for our sense of importance as a species? Do we become more self-assured and, as a society, act more like the guardians of this universe of ours? Or do we wallow in pity and lack of direction knowing we might be some kind of 'cosmic mistake'?

And if we find out we are not alone, how do we sit with the knowledge there are others out there but that we're unlikely to get to communicate with them in our lifetime? And how would we react to either a possible threat or a possible friend?

Finding life elsewhere isn't just about trying to find out what's 'out there'; it's about understanding the context of life in the universe. That means learning more about life here on Earth just as much as that beyond it. The questions we ask can easily advance from 'How might life survive in other places?' to 'Why might our civilisation survive when others don't, or what might endanger our survival?'

This field is both about finding life elsewhere as well as finding ourselves: 'Are we alone?' is intrinsically tied to 'Why am I here?'

THE FRAGILE QUESTIONS WE'RE SCARED TO PONDER

When we get lost in the hype around aliens and UFOs, and then tie them to the broader search for life elsewhere, we somehow manage to forget the reality in which we live. When you put the pseudoscience to one side, you find you're asking fundamental questions about why we're here, how life happened, and the general nature of our universe. It's undeniable that we are each here, living our short lives on this little planet; wondering why and how is not daft, they are understandable questions to be asking. In some sense, if you have those moments every now and again where you stop and consider our lack of knowledge of what life is all about, it seems quite strange to not indulge in pondering what the answer might be. Why *wouldn't* you think to ask?

When hype and fear take over, the loser is society as a whole, in that the bigger questions that arise from an open-minded, rational line of questioning about extraterrestrial life aren't pondered by the many. There are the scientific questions, of course, but there are also the more internal questions, the more human questions, that cannot be avoided when you take seriously the idea of life beyond planet Earth, that everyone has a right, and a level of responsibility, to ask. These are questions we cannot be scared to ask for fear of veering into pseudoscience, or for fear of being seen as pretentious and with our 'heads in the clouds'. Genuinely asking the questions 'Are we alone?' and 'Why are we here?' is not unrealistic or pointless or strange; they are the most self-reflective questions we can possibly ask, and no one should feel unqualified to consider them.

If we think critically and aren't too proud to update our thinking with new evidence, whether that's in the form of a basic article, an opinion of an expert we find to be reliable or a scientific paper itself, the world of science and the fascinating unanswered questions it brings become much more open to those who feel they're on the outside. Of course, educating ourselves as to what counts as a reliable or unreliable source, and what constitutes a strong argument over a weak one, is vital if we're to ponder these questions effectively, but if we succumb to the fear of 'getting it wrong' or assuming these questions are only asked by fringe members of society, we're all missing out.

We're not just missing out on wonderful, awe-inspiring, fundamentally bigger-than-thou thinking, though. There's an argument to be made that when society doesn't ask itself these momentous questions, it fails to see past its immediate period of time and physical surroundings.

BEYOND CURIOSITY

Maybe we'll never know the answer to whether we're alone, but maybe the quest to answer it might have far more immediate effects than working out who wins the cosmic bet or pondering our very existence.

In 2018, Professor Ian Crawford of Birkbeck, University of London, published a paper titled: 'Widening perspectives: the intellectual and social benefits of astrobiology (regardless of whether extraterrestrial life is discovered or not)'.[35] In it, he argued that astrobiology does a grand job of pulling together scientists from across the disciplines of astronomy, biology, chemistry, geology and planetary science, stimulating a reunification of the sciences from the twenty-first-century siloed approach to that of the interdisciplinary efforts of the seventeenth and eighteenth centuries. He writes: 'It is from this cross-fertilisation of ideas that future discoveries that would not otherwise be made may be expected, and such discoveries will comprise a permanent legacy of astrobiology even if they do not include the discovery of alien life.'

This is not Crawford's full argument, though. Echoing the sentiment of Carl Sagan in his famous 1994 *Pale Blue Dot*, Crawford argues that simply by considering the solar system we inhabit, the enormity of our planet, the impossible-to-truly-comprehend age of our universe, people are forced to think beyond the limits of their immediate existence.

Crawford calls this 'cosmic perspective', which reveals Earth to be a very small planet adrift in the universe but, to date, the only known inhabited place within it. He also speaks about the 'evolutionary perspective', which reveals that all life on Earth is related, due to its common origin and evolutionary history. These perspectives might seem obvious, but they are indeed often lost in the narrow nationalistic and unscientific ideologies that still capture many people today. Crawford argues that prompting more people to consider the questions astrobiologists and SETI researchers ask can only result in a society with a far more expanded, and thus hopefully more educated and well-rounded, worldview.

In 1965, just four years after Yuri Gagarin became the first human being to travel in space, the US Ambassador to the United Nations, Adlai Stevenson, addressed the UN with this: 'Just as Europe could never again be the old closed-in community after the voyages of Columbus we can never again be a squabbling band of nations before the awful majesty of outer space.'[36] Beyond better geopolitical relations, though, with more people actively considering the cosmic and evolutionary perspectives, there's an argument to be made that many other socially beneficial considerations might be prioritised in the minds of the many. For example, when you consider that we might be the only form of life in the universe and that the habitability of the Earth is decreasing due to climate change, the draining of resources and the destructive expansion of human societies into crucial ecological habitats, moral responsibility around simple individual actions could be heightened.

WHY WE MUST THINK BIGGER AND BETTER

'Zooming out' seems to have been a bit of a thread throughout this book, and I hope that I've convinced you that thinking about industries, claims and questions through a bigger lens is crucial to cementing understanding and finding nuance. But with the search for life elsewhere, zooming out not only allows us to consider big questions about the nature of the universe and who we are as human beings, but also gives us a tool in reframing how we consider questions, perspectives and opinions about how the world works in general. That tool of perspective, of humble scepticism and thinking beyond your own immediate situation, is really how we cut through hype, and come to terms with the nuance, the conflicting ideas and the complex systems that power our society.

If we can think beyond our desire for technology, maybe we can better uncover the problematic flaws in the systems that provide for us, and, as a society, do better. If we can think beyond our emotional and financial attachments to things, maybe we can see where these systems are lacking for the many and pressure those in power to make them fairer. If we can step back from our immediate reactions to new ideas, and work out whether those reactions are down to human nature or external, perverse incentives, maybe we can see beyond the smoke and mirrors distracting us from the bigger picture.

We have the ability, as human beings with imaginative minds, to think about problems from many different angles, timeframes, geographies and cosmic distances. And there's a responsibility too that comes with this superpower of sorts, and that is 'critical thinking'. If astrobiology and the search for life elsewhere help to instil that global consciousness for considering things beyond our own singular frame of reference, then we cannot afford to silo them off to only the few who are currently willing to think beyond the pseudoscience.

Conclusion

....

There's one piece of hype that, when believed, serves to enhance and propagate all the other hype in the science and technology world, and it is this: 'I'm not a scientist, or trained in this field, or someone who was good at maths at school, so it is not my place to question, to engage, or to look into this further.'

Science and technology is seen as this 'other' thing that only the clever people 'get'; that only the trained can fully appreciate; and only those with either huge amounts of time to spare or knowledge already in their pockets could ever possibly challenge or put into context what maybe, just maybe, the experts might have got wrong.

Science as the 'ivory tower' is a narrative that is propagated less in the headlines and more in the insecurities, historical narratives and individual fear in each and every one of us. It grows over time in many people, so much so that by the time they're adults, they believe themselves to be incapable of engaging with a better understanding of our world.

I hope in these pages I've shown both how much of a pity believing *that* hype is, but also how problematic and dangerous it can be when too many buy into this 'it's not my place' narrative. We cannot succumb to our perceived inability to work stuff out.

Critical thinking is not something intelligent people are born with. It's a skill they have practised by asking questions, both of others and of themselves, while sitting comfortably in a state of ignorance. Critical thinking is not simple, but it's certainly not reserved for those who started young.

For people who like 'how-to' guides, a nine-step plan for beating the hype, one for each chapter, would be a good way to end this book. And if I was to consider what that might look like, maybe it would read as follows:

1. *Hype maintains the status quo:* so work out what the deeper issues are that currently exist in the world, and question whether change-based claims really do solve those real problems.
2. *Hype is a double-edged sword:* so consider how excitement around an idea might be wielded for both good and bad.
3. *Hype shields complexity:* so map the full system, find the conflicting balancing act within, read around, and consider how things play out for different people, in different countries, in different timeframes.
4. *Hype curbs action:* so work out what thoughts or fears or cultural tendencies are stopping people from doing.
5. *Hype shields flaws:* so go in assuming everything has nuance, and that absolutist statements miss out key points.
6. *Hype is a placebo:* so consider who wins by having the masses believe.
7. *Hype is fanatical:* so question emotional desire versus societal need.
8. *Hype relinquishes responsibility:* so consider who else is in control if you are not, and the decision-making power that this control provides.
9. *Hype halts rational thinking:* so work out exactly how 'being a human' is working against you, whether that's your cognitive biases, your fears, your culture, your dogged pursuit of status or your natural fallibility.

Unfortunately, though, critical thinking isn't just about pulling out a checklist of standardised questions to run by any statement we might come across.

Understanding the world means doing the hard task of pushing away the comfortable illusion that falsely reassures us, or gives us too-good-to-be-true answers, or tells us that we're good people who don't need to change. This is what makes critical thinking, and beating hype, so hard; we have to consider for a moment that we ourselves are complicit in the problematic outcomes they might bring.

So, to be better critical thinkers, there are a few things we must first humbly accept: that there is no simple formula to quickly understand the world; that we can never know anything in its entirety; and that we must consider our own failings, our own prior judgements and the effect our own behaviour has on the world around us . . . and stare them all straight in the face.

We have to accept that the smoke and mirrors might not have been assembled by someone else, but in fact by ourselves.

My hope in writing this book isn't to scare, blame or guilt-trip. My hope is that in better understanding hype and knowing that there are always questions to be asked, we will all feel more empowered to try and see past it. Questioning hype isn't about offering the correct answer, it's an act of confirming, of testing, of putting into other contexts to see if it still holds water; it's about being open to claims being right or wrong, not assuming one side and seeking to confirm; it's simply the reasonable, logical thing to do when faced with new information, and the world needs more of us to do it.

No one is immune from being swept up in media frenzy, or ideas that feel so believable on the surface, no matter how expert they might be in a particular topic. It's impossible to have deep knowledge about every facet of society affected by a particular area of science and technology, and thus it is easy to be mistaken, whether you have a PhD or simply a passing interest. There are always gaps in knowledge. Only the uneducated always feel right; everyone else is second guessing themselves on a daily basis. The empowered among us welcome their own ignorance, and use that unmasking of themselves as unknowing as a tool to then learn more. They open the floodgates of 'Oh God, where to start?' and use their trusty, humble 'So why is that?' dinghy to tackle the treacherous waves.

Those disempowered by the idea of not knowing the right answer hide their ignorance within themselves and therefore, ironically, their ability to get rid of it. They miss out on the joy, wonders and fulfilment of the extreme sport called learning.

Critical thinking isn't about being negative, but about being open to the idea that something might be wrong. And that openness, which ultimately leads to actively engaging in the topic, can be far more interesting than simply being fed entertainment in the form of tech-hype porn. Understanding the world doesn't only mean seeing its dark side, it also means revelling in its complexity, and the realisation that we live on a planet so rich in open questions that we will never find an end to discovering something new. And it's about getting closer to understanding why the things that don't work, don't work, and therefore find hints as to how to fix them.

That's empowering, not depressing.

There's one final point that has been at the root of every chapter, and is indeed at the root of everything around us, and that's our own level of engagement.

We cannot possibly unearth every deeper truth and act on every new piece of information we come across; that surely isn't what life is about. But what we must realise is that we have a choice, a difficult choice. And that is how we choose to react to what's behind hype. As we've discovered, hype often provides a comfortable illusion; so what does it mean to have that illusion removed and 'see the light'? De-blinkering ourselves means asking complex questions such as 'What is right and wrong?', 'What is more important?' and 'Am I willing to change?', and these questions are easier to ignore than to try to wrestle with.

It's worth pondering if we really do want to see past the hype. Or rather, now you've read this book, how at peace you might be with actively not knowing.

Being blinkered in our view of the world doesn't just make us ignorant or neutral in our behaviour; that distraction actively plays into a status quo that quite often is having a negative overall effect on the world around us. Believing hype without question sometimes means we're inadvertently promoting, encouraging and taking part in something not right, making us a bystander of sorts.

It's not enough to just take a surface level consideration of the sciences; and it's not enough to follow along and delight at the tech-innovation headlines; we must understand the role they play in society as a whole, and in the way they shape geopolitical systems, the environment, socioeconomics and culture. We don't all need to know the nitty-gritty details, that's the job of the academics, but we all have a responsibility to not take science and technology at face value, and blindly believe whatever is trending. It's not good for us.

H. G. Wells wrote in his 1920 *Outline of History*: 'human history becomes more and more a race between education and catastrophe'.[1] I believe education isn't just about knowing things in great detail, it's about nuance and context and consideration of how different things interact with one another. And if we collectively want to improve the state of the world, as opposed to simply inhabit it, we cannot get lost in the hype that surrounds certain siloed ideas. Understanding how the world works isn't made easy by the way news is shared, granted. But the cost of not critically thinking, of taking the easy route in not questioning the world around us, is too high to put down to fear, difficulty, or lack of time or care.

Critical thinking is what unearths the roots of problems, it is what finds understanding and empathy in conflicting views, and it is what allows for society collectively to come to more logical conclusions. And if you believe, as I do, that if everyone were to think more thoroughly, they'd also think more altruistically and fairly, too, then critical thinking really could be something we can all actively do to enhance the state of our world, right there from where you're sitting or lying or standing or running or walking this very second.

Allow me then, with the greatest respect, to update Wells's famous line: 'humanity's future becomes more and more a race between critical thinking and catastrophe'.

Acknowledgements

To so many pals who listened to me bang on about this book over lunches, dinners and gins, and gave me feedback, food for thought and endless encouragement: cheers, guys.

To my writer pals Phil, Alexa and Emma in particular, who nerded out with me about structure and themes and titles (oh god, the title . . .), and all the more pernickety stuff you were happy (I think?) to listen to me waffle on about: thank you so much.

To my incredible beta readers Bec, Charly, Claire, Emma, Harvey, Hila, Joe, Kitty, Lauren, Maria, Martina, Maxine, Roman and Susi, who not only gave up their precious time to read my words, but were all so compassionate when telling me which of those words didn't work: this book is better because of all of you.

To my wonderful friend Parul, who introduced me to the wonderful world of the Story Grid, and gave me so much confidence at the start when, frankly, I was freaking out at the prospect of writing something longer than an article: thank you.

To the lovely Leigh, who helped me turn my first draft into something legible: you are a magician.

To my agent Laura, who took a punt on me and has been very patient with my never-ending stream of questions ever since: I'm so grateful, pal.

To all those at Robinson at Little, Brown who've played a part in bringing this book into the world: thank you so much for helping me make this a reality.

To my brother Gordon, who is always the most generous and support-ive cheerleader: thank you for your endless woops in our family's WhatsApp group.

To my brother Ross: you may not believe it yourself, but you're an inspiration to the rest of us.

To Dad, my biggest inspiration of them all: the part you've played in making me, and encouraging me to be, the stubborn science nerd I am is in no way insignificant.

To Mum, my best pal, who has been at the other end of the phone for all of my editorial and emotional needs: there really aren't the words to say thank you enough.

And, of course, to Lawrence. Who listened to so many brainstorms and rambles; who shared his own abundant expertise; who read (almost) the whole book and told me all his sometimes-too-honest thoughts; who brought me Magic Stars and Irn Bru when I thought I couldn't keep writing; and generally kept me sane throughout this whole surreal process: I'm forever grateful that you rocked up into my life.

* * *

To the wonderful experts who contributed their time and expertise in inter-views, chapter reviews and comments – thank you so much:

Feeding the World
Abby Schlageter, Farmarama
Bernhard Kowatsch, World Food Programme
Emma Cosgrove, Supply Chain Dive

Curing Cancer
Charlotte Casebourne, Theolytics
Eilish Middlehurst & Matthew Griffiths, ConcR
George Tetley, Deep Science Ventures
Justin Stebbing, Imperial College London
Kat Arney, Science Writer

Lawrence Yolland, University College London
Rebecca Newton, recovered cancer patient

Fuelling All Our Tech
Akshat Rathi, Bloomberg
Alice Bell, 10:10
Enass Abo-Hamed, H2GO Power
Simon Engelke, Cambridge University

Limitless Energy
Akshat Rathi, Bloomberg
Andrew Steele, Scientist & Writer
Ian Chapman, UK Atomic Energy Authority
Malcolm Handley, Angel Investor

Commmercialising Space
Daniel Faber, Orbit Fab
Harriet Brettle, Astroscale
Jeff Hill, NYU
Manny Shar, Bryce Space and Technology
Mark Boggett, Seraphim Space
Paul Liias, Estonian Ministry of Economic Affairs
Thomas Cheney, SGAC Space Law and Policy

Understanding the Universe
Christophe Jurczak, Quantonation
Frank Ruess, Google
Martin Laforest, Institute for Quantum Computing, University of
    Waterloo
Philipp Gerbert, The Boston Consulting Group
Scott Aaronson, University of Texas at Austin
Theodor Lundberg, Cambridge University

Just Think It
Anne-Laure Le Cunff, King's College London

Oliver Armitage, BIOS
Timothy Constandinou, Imperial College London

Decision-making on Tap
Ceyhun Karasu, Independent AI Researcher
James Kingston, CogX
Jeremy Waite, IBM
Joanna Bryson, Hertie School of Governance

Finding Life Elsewhere
Brad McLain, National Centre for Women and Information Technology
Graham Lau, Astrobiologist & Science Communicator
Greg Johnson, former Astronaut
Lewis Dartnell, University of Westminster

Hype Themes
Anthony Owen, Magician & Producer
Christine Aicardi, King's College London
Dawn Walter, Anthropologist
Ellen Anthoni, Trend researcher
Elsa Sotiriadis, Biofuturist
Helen Smith, Independent researcher
Hannah Pitt, RBS
Maria Jeansson, Imperial College London
Orowa Sikder, Oxford University
Tim Stroh, Author of 'A Deeper Truth'

Notes

....

INTRODUCTION

1. Edme Goyet, *Nouvelles récréations physiques et mathématiques*, Second Edition, Paris, 1786, p.240.

FEEEDING THE WORLD

1. A. de Groot Ruiz et al., 'The external costs of banana production: A global study', Fairtrade International / True Price / Trucost, 2017, maxhavelaar-uploads.s3.amazonaws.com/uploads/download_item_file/file/151/170224_Research_Report_External_Cost_of_Bananas_-_final.pdf
2. Joe Fassler, 'Bananas are getting cheaper. That low price comes with hidden costs', *New Food Economy*, May 20, 2019.
3. Data has been collected between 2013 and 2018 by Euromonitor and presented by the USDA, listing 86 countries for which this information is available. The UK is third to last in this list, ers.usda.gov/topics/international-markets-us-trade/international-consumer-and-food-industry-trends.
4. Simon Gompertz, 'How did households budget in 1957?', BBC, January 18, 2018.
5. Mark Solomons, 'Price of milk soars as farmers feels effect of summer heatwave', *The Express*, November 16, 2018; Kate Buck, 'Supermarket milk prices are about to go up 7.5%', *Metro*, November 15, 2018.
6. 'Tesco removes Marmite and other Unilever brands in price row', BBC, 13 October, 2016.

7. Madeleine Howell and Gareth May, 'The hidden cruelty of the cashew industry – and the other fashionable foods that aren't as virtuous as they appear', *The Telegraph*, April 4, 2019.

8. Emma Mills, 'Why your avocado toast could be destroying Mexican forests', *The Telegraph*, August 17, 2016.

9. Kendra Pierre-Louis, 'No one is taking your hamburgers. But would it even be a good idea?', *The New York Times*, March 8, 2019.

10. The 70% result comes from the World Bank's 'World Development Indicators', as collected in 2017; this statistic is summarised and can be interactively explored at blogs.worldbank.org/opendata/chart-globally-70-freshwater-used-agriculture.

11. European Environment Agency, 'Water for agriculture', July 3, 2012, last modified August 31, 2016, eea.europa.eu/articles/water-for-agriculture.

12. Elizabeth Curmi et al., 'Feeding the future: How innovation and shifting consumer preferences can help feed a growing planet', November, 2018, citibank.com/commercialbank/insights/assets/docs/2018/feeding-the-future.pdf.

13. *Ibid.*

14. Food and Agriculture Organization of the United Nations (FAO), 'The state of the world's biodiversity for food and agriculture', 2019, fao.org/3/CA3129EN/ca3129en.pdf. The FAO have created this great interactive version of their report, and an associated podcast, which I'd recommend exploring, and can be accessed here fao.org/state-of-biodiversity-for-food-agriculture.

15. Clive Cookson, 'Scientists see role for insects and "orphan crops" in human diet', *Financial Times*, December 7, 2017.

16. *See note 12* (Curmi et al).

17. Amy R Beaudreault, 'Lab to fork: A lost opportunity for global nutrition', *Center for Strategic and International Studies*, May 2, 2019.

18. Afshin *et al.*, 'Health effects of dietary risks in 195 countries, 1990–2017: a systematic analysis for the Global Burden of Disease Study 2017', *The Lancet*, 2019, 393(10184), 1958-72. DOI: 10.1016/S0140-6736(19)30041-8.

19. *See note 17* (Beaudreault).

20. These numbers are taken from a summary of the key findings from the FAO's 'Save Food: Global Initiative on Food Loss and Waste Reduction', updated in 2019, and can be found at fao.org/save-food/resources/keyfindings.

21. *See note 12* (Curmi et al).

22. Amy R Beaudreault, 'Nutrition Policy Primer: The Untapped Path to Global Health, Economic Growth, and Human Security', March 2019, csis.org/features/nutrition-prosperity.
23. Funding information for the World Food Programme can be found at wfp.org/funding-and-donors, accessed July 2019.
24. Christopher J Elias, 'Why working with farmers is key to fighting hunger', *Agenda* (a publication of the *World Economic Forum*), August 26, 2015.
25. Max Kutner, 'Death on the farm', *Newsweek*, April 10, 2014. This was an investigative piece focusing specifically on suicide in farming. There is much discussion about the link between suicide and the unpredictability of the farming profession worldwide, and the reporting in this space is complex due to sometimes inaccurate self-reporting of professions. The research continues to evolve.
26. Tim Lang & Victoria Schoen, 'Britain's love affair with cheap food could be coming to an end', *The Conversation*, August 5, 2014.
27. Patrick Butler, 'More than a million UK residents live in "food deserts", says study', *Guardian*, October 12, 2018. The study itself was conducted by the Social Market Foundation in 2018, funded by Kellogg's; I've referenced the *Guardian* piece to give a broader context.
28. Susie Quick, 'A town called Malnourished', *Newsweek*, April 3, 2014. They have cited the US Department of Agriculture for their data.
29. Michael Pooler, 'Retailers, distributors and growers struggle to curb food waste', *Financial Times*, December 7, 2017. They have cited the FAO for their data.
30. *See note 26*. See also Tim Lang, Erik Millstone and Terry Marsden's research into the potential effects of Brexit on the UK food system: 'A food Brexit: time to get real – A Brexit briefing', accessed December 2019, can be found at http://openaccess.city.ac.uk/id/eprint/18655/.
31. *AgFunder*, 'AgFunder AgriFood Tech Investing Report', 2018, the report can be downloaded for free (once you put in a name and email address) at agfunder.com/research/agrifood-tech-investing-report-2018/ (as of December 2019).
32. *World Economic Forum*, 'Meat: the future series - Alternative proteins', January 2019, weforum.org/docs/WEF_White_Paper_Alternative_Proteins.pdf
33. Soutik Biswas, 'The myth of the Indian vegetarian nation', BBC, April 4, 2018.
34. 'World's first lab-grown burger is eaten in London', BBC, August 5, 2013.
35. *See note 15* (Cookson).
36. Jonathan Smith, 'French Company to Open Biggest Insect Farm in the World', *Labiotech*, February 22, 2019.

37. Marie Mawad, 'Giant French insect farm managed by robots wins new investment', *Bloomberg*, February 20, 2019.

38. John Lynch & Raymond Pierrehumbert, 'Climate impacts of cultured meat and beef cattle', *Front. Sustain. Food Syst.*, 2019, 3(5). DOI: 10.3389/fsufs.2019.00005.

39. *See note 32* (WEF).

40. Joe Andrews, 'Beyond Meat, Impossible Foods and the plant-based burger of the summer. Here are the diet facts you need to know', CNBC, July 4, 2019.

41. American Association for the Advancement of Science, 'Statement by the AAAS Board of Directors On Labeling of Genetically Modified Foods', October 20, 2012; The American Medical Association position statement can be accessed here (as of December 2019): https://www.isaaa.org/kc/publications/htm/articles/position/ama.htm; an EU report that discussed GMO research between the years 2001 and 2010 can be found here: https://ec.europa.eu/research/biosociety/pdf/a_decade_of_eu-funded_gmo_research.pdf; Jane E Brody, 'Are G.M.O foods safe?', *The New York Times*, April 23, 2018; 'Fields of gold', *Nature*, May 2, 2013, 497, 5-6. DOI: 10.1038/497005b; Amy Harmon, 'A lonely quest for facts on Genetically Modified Crops', *The New York Times*, January 4, 2014; Megan L Norris, 'Will GMOs hurt my body? The public's concerns and how scientists have addressed them', Harvard University's *Science in the News*, August 10, 2015.

42. Jane E. Brody, 'Are G.M.O. foods safe?', *The New York Times*, April 23, 2018.

43. Tom Colicchio, 'Are you eating Frankenfish?', *The New York Times*, December 15, 2015.

44. Philip M Fernbach et al., 'Extreme opponents of genetically modified foods know the least but think they know the most', *Nature Human Behaviour*, 2019, 3, 251-256. DOI: 10.1038/s41562-018-0520-3.

45. Ian Sample, 'Strongest opponents of GM foods know the least but think they know the most', *Guardian*, January 14, 2019.

46. *Confédération paysanne and Others v Premier ministre and Ministre de l'Agriculture, de l'Agroalimentaire et de la Forêt* (2018) Case C-528/16, Court of Justice of the European Union.

47. Jessica Pothering, 'Europe's gene editing regulation exposes the messy relationship between science and politics', *AgFunderNews*, April 10, 2019.

48. International Fund for Agricultural Development, 'Smallholders, food security, and the environment', 2013, ifad.org/documents/38714170/39135645/smallholders_report.pdf.

49. FAO, 'Food security for sustainable development and urbanization: Inputs for FAO's contribution to the 2014 ECOSOC Integration Segment,

27-29 May', 2014, un.org/en/ecosoc/integration/pdf/foodandagricultu-reorganization.pdf.

50. Lydia Mulvany, Kevin Varley, & Cedric Sam, 'The changing face of farms: Women step in as U.S. growers age', *Bloomberg*, April 11 2019; Torsten Kurth *et al.*, 'It's time to plant the seeds of sustainable growth in agriculture', BCG, September 4, 2018.

51. Andrew Jenkins, 'Food security: vertical farming sounds fantastic until you consider its energy use', *The Conversation*, September 10, 2018.

52. Jonathan Foley, 'Local food is great, but can it go too far?', *GreenBiz*, October 4, 2016.

53. Christopher Mims, 'Are shipping containers the future of farming?', *Wall Street Journal*, June 8, 2016.

54. Stan Cox, 'Enough with the vertical farming fantasies: There are still too many unanswered questions about the trendy practice', *Salon*, February 17, 2016.

55. ReportsnReports, 'Vertical Farming Market by Growth Mechanism (Hydroponics, Aeroponics, and Aquaponics), Structure (Building Based and Shipping Container), Offering (Hardware, Software, and Service), Crop Type, and Geography – Global Forecast to 2022', January 2017.

56. Hamish Cunningham & Benz Kotzen, 'The sustainable vegetables that thrive on a diet of fish poo', *The Conversation*, November 10, 2015.

57. Andrew Jenkins, 'Why vertical farming isn't a miracle solution to food security', *Independent*, September 28, 2018.

58. *See note 31* (*AgFunder*). Have summed the 2018 investments in Cresco Labs, Agricool and Infarm on page 30. It should be noted that Cresco Labs, with the biggest investment of $100 million, is a cannabis growing vertical farm – make of that what you will.

59. *Ibid.*

60. George Monbiot, 'The shocking waste of cash even leavers won't condemn', *Guardian*, June 21, 2016.

61. Department of Environment Food and Rural Affairs, 'The new Common Agricultural Policy schemes in England: October 2014 update', 2014.

62. Fiona Harvey, 'Big farmers to see funding cut post-Brexit after Gove shakeup', *Guardian*, September 12, 2018.

63. Environmental Working Group's Farm Subsidy Database, accessed December 2019, can be found at this address with these search terms: farm.ewg.org/progdetail.php?fips=00000&progcode=totalfarm&page=conc&regionname=theUnitedStates.

64. United States Department of Agriculture, 'America's Diverse Family Farms, 2018 Edition', December 2018, ers.usda.gov/webdocs/publications/90985/eib-203.pdf.

65. Robert Coleman, 'The rich get richer: 50 billionaires got federal farm subsidies,' *Environmental Working Group*, April 18, 2016. The EWG looked at the period between 1995–2014.

66. Rural Payments Agency, 'The Basic Payment Scheme in England', 2015; *see also note 61*.

67. Environmental Working Group's Farm Subsidy Database, accessed December 2019, can be found at this address with these search terms: farm.ewg.org/progdetail.php?fips=00000&progcode=total_dp&regionname=theUnitedStates.

68. Jonathan Foley, 'It's time to rethink America's corn system', *Scientific American*, March 5, 2013.

69. Kimberley Amadeo, 'Farm subsidies with pros, cons, and impact', *The Balance*, updated July 4, 2019, accessed December 2019.

70. Siegel *et al.*, 'Association of higher consumption of foods derived from subsidized commodities with adverse cardiometabolic risk among US adults', *JAMA Internal Medicine*. 2016, 176(8), 1124-1132. DOI:10.1001/jamainternmed.2016.2410.

71. U.S. Department of Health and Human Services & U.S. Department of Agriculture, 'Dietary Guidelines for Americans 2015-2020: Eighth edition', accessed December 2019, health.gov/dietaryguidelines/2015/guidelines/.

72. Gabe Brown, *Dirt to soil: One family's journey into regenerative agriculture*, Chelsea Green Publishing, 2018, pp.179–182.

73. *Ibid.*

74. Lara Bryant, 'Organic matter can improve your soil's water holding capacity', Natural Resources Defense Council, May 27, 2015; Emily Payne, 'Regenerative agriculture is getting more mainstream but how scalable is it?', *AgFunder*, May 28, 2019.

75. Jessica McKenzie, 'Regenerative agriculture could save soil, water, and the climate. Here's how the U.S. government actively discourages it', *The New Food Economy*, March 14, 2019.

76. Wayne Arnold, 'Surviving without subsidies', *The New York Times*, August 2, 2007.

77. Tony St Clair, 'Farming without subsidies – a better way. Why New Zealand agriculture is a world leader', *Politico*, August 17, 2002.

78. Robert Costanza et al., 'Development: Time to leave GDP behind', *Nature*, 2014, 505, 283-285. DOI: 10.1038/505283a.

79. Mariana Mazzucato, *The value of everything: Making and taking in the global economy*, Allen Lane, 2018.

80. *See note 78* (Costanza et al.).

81. Ida Kubiszewski *et al.*, 'Beyond GDP: Measuring and achieving global

genuine progress', *Ecological Economics*, 2013, 93, 57-68. DOI: 10.1016/j.ecolecon.2013.04.019.

82. Pamela Mason & Tim Lang, *Sustainable Diets*, Routledge, 2017.

FUELLING ALL OUR TECH

1. Tasha Robinson, 'Modern horror films are finding their scares in dead phone batteries', *The Verge*, August 16, 2018.

2. Robert Rapier, 'The Holy Grail of lithium batteries', *Forbes*, May 16, 2019; 'Patrick Soon-Shiong on 'Holy Grail' zinc-air battery', CNBC Video, September 26, 2018; 'Tesla's electric vehicle Holy Grail', *AllThingsEV.info*, April 10, 2019; Peter Maloney, 'Chasing the Holy Grail of battery storage, scientists test solid state magnesium electrolyte', *Utility Dive*, December 19, 2017; Kent Moors, 'MIT may have just given us the "Holy Grail" of battery breakthroughs', October 19, 2017.

3. *Bloomberg*, 'Electric Vehicle Outlook 2019', 2019, full report can be downloaded at about.bnef.com/electric-vehicle-outlook/; Mark Kane, 'Global Sales December & 2018: 2 million plug-in electric cars sold', *InsideEVs.com*, January 31, 2019.

4. Ivan Penn, 'California invested heavily in solar power. Now there's so much that other states are sometimes paid to take it', June 22, 2017.

5. Jeremy Berke, 'Germany paid people to use electricity over the holidays because its grid is so clean', *Business Insider*, December 29, 2017.

6. REN21, 'Renewables 2019: Global status report', 2019, ren21.net/wp-content/uploads/2019/05/gsr_2019_full_report_en.pdf.

7. Akshat Rathi, 'Why energy storage is key to a global climate breakthrough', *Quartz*, June 5, 2018.

8. Companies taking these approaches include Sila Nanotechnologies, Enovix, QuantumScape & Solid Power.

9. Companies taking these approaches include Teraloop, Brenmiller Energy, H2Go Power & Quidnet Energy.

10. Logan Goldie-Scot, 'A behind the scenes take on lithium-ion battery prices', *BloombergNEF*, March 5, 2019.

11. Akshat Rathi, 'How we get to the next big battery breakthrough', *Quartz*, April 8, 2019.

12. Matthew R. Shaner et al., 'Geophysical constraints on the reliability of solar and wind power in the United States', *Energy & Environmental Science*, 2018, 11, 914–925. DOI: 10.1039/c7ee03029k.

13. Fernando J.de Sisternes et al., 'The value of energy storage in decarbon-izing the electricity sector', *Applied Energy*, 2016, 175, 368-379. DOI: 10.1016/j.apenergy.2016.05.014.

14. James Temple, 'Relying on renewables alone significantly inflates the cost of overhauling energy', *MIT Technology Review*, February 26, 2018; James Temple, 'The $2.5 trillion reason we can't rely on batteries to clean up the grid', *MIT Technology Review*, July 27, 2018; Jason Pontin, 'We gotta get a better battery. But how?', *Wired*, September 17, 2018.

15. Akshat Rathi, 'To hit climate goals, Bill Gates and his billionaire friends are betting on energy storage', *Quartz*, June 12, 2018.

16. This section owes a lot to the brilliant reporting in the following deep dives on BYD: Akshat Rathi & Echo Huang, 'Beyond the Tesla bubble: The future of electric cars is being scripted in China', *Quartz*, December 10, 2018; Akshat Rathi & Echo Huang, 'Inside BYD—the world's largest maker of electric vehicles', December 13, 2018; Matthew Campbell & Ying Tian, 'The world's biggest electric vehicle company looks nothing like Tesla', *Bloomberg Businessweek*, April 16, 2019.

17. 'China is leading the world to an electric car future', *Bloomberg Businessweek*, November 15, 2018; IEA, 'Global EV Outlook 2018', May 2018; chart of EV sales from IEA report created and shared by Akshat Rathi and can be found at theatlas.com/charts/HkDEZJVA7.

18. Scott Kennedy, 'China's expensive gamble on new-energy vehicles', *Center for Strategic and International Studies*, November 6, 2018.

19. 'In China, the License Plates Can Cost More Than the Car', *Bloomberg News*, April 25, 2013.

20. Yuyu Chen et al., 'Evidence on the impact of sustained exposure to air pollution on life expectancy from China's Huai River policy', *PNAS*, 2013, 110(32), 12936-12941. DOI: 10.1073/pnas.1300018110; *also see note 17* (*Bloomberg Businessweek*).

21. 'Volkswagen opening three new factories in China to handle SUV, elec-tric boom', *Bloomberg News*, May 28, 2018.

22. 'Elon Musk in China to break ground for Tesla's China factory', *Bloomberg News*, January 7, 2019.

23. Kate O'Flaherty, 'Huawei security scandal: Everything you need to know', *Forbes*, February 26, 2019.

24. This section owes a lot to the brilliant reporting in the following deep dive on CATL: Akshat Rathi, 'The inside story of how CATL became the world's largest electric-vehicle battery company', *Quartz*, April 3, 2019.

25. *Ibid.*; 'Valmet Automotive and CATL form a strategic partnership in elec-tric vehicle solutions – CATL invests in Valmet Automotive to become an important owner', January 30, 2017; Jonathan Shieber, 'Volvo inks

multi-billion-dollar battery deals with LG Chem and CATL for EVs', May 15, 2019.

26. *See note 24* (Rathi).
27. Akshat Rathi, 'Five things to know about China's electric-car boom', *Quartz*, December 2018.
28. *See note 16* (Rathi, December 13, 2018).
29. Andrew J Hawkins, 'Bird has a new electric scooter: it's durable, comes in three different colors, and you can buy it', *The Verge*, May 8, 2019.
30. Derek Watkins, K.K. Rebecca Lai & Keith Bradsher, 'The world, built by China', *The New York Times*, November 18, 2018.
31. 'Our bulldozers, our rules', *The Economist*, July 2, 2016.
32. *See note 30* (Watkins et al).
33. Maria Ab-Habib, 'How China got Sri Lanka to cough up a port', *The New York Times*, June 25, 2018; Peter S. Goodman & Jane Perlez, 'Money and muscle pave China's way to global power', *The New York Times*, November 25, 2018.
34. Jason Horowitz & Liz Alderman, 'Chastised by E.U., a resentful Greece embraces China's cash and interests', *The New York Times*, August 26, 2017.
35. *See note 30* (Watkins et al).
36. Todd C. Frankel, 'The cobalt pipeline', *Washington Post*, September 30, 2016.
37. *Ibid.*
38. 'U.N. points finger at "elements" in Congo army over Kasai mass graves', *Reuters*, July 25, 2017.
39. David Fickling, 'Cobalt's chemistry experiment', *Bloomberg Opinion*, September 28, 2017.
40. Benchmark Mineral Intelligence, 'Panasonic reduces Tesla's cobalt consumption by 60% in 6 years... but cobalt supply challenges remain', *The Assay*, June 6, 2018.
41. Eshe Nelson, 'The hottest thing in the markets right now is an obscure metal mined in DR Congo', *Quartz*, April 27, 2017.
42. European Commission guidance on conflict materials, accessed December 2019, can be found at ec.europa.eu/trade/policy/in-focus/conflict-minerals-regulation/regulation-explained/index_en.htm; Global Witness briefing on US conflict minerals law Section 1502 of the US Dodd Frank Act, accessed December 2019, can be found at globalwitness.org/en/campaigns/conflict-minerals/dodd-frank-act-section-1502/.
43. Akshat Rathi, 'One Chinese company now controls most of the metal needed to make the world's advanced batteries', *Quartz*, May 30, 2018.
44. Keith Bradsher, 'China is blocking minerals, executives say', *The New York Times*, September 23, 2010.
45. Ashley Feng & Sagatom Saha, 'Chinese heavy metal: How Beijing could

use Rare Earths to outplay America', *Scientific American*, August 3, 2018; 'Fears rise China could choke off supply of Rare Earths in trade war', *Bloomberg Businessweek*, May 23, 2019.

46. David Barboza, 'China unveils $586 billion stimulus plan', *The New York Times*, November 10, 2008.
47. Keith Bradsher & Li Yuan, 'China's economy became No. 2 by defying No. 1', *The New York Times*, November 25, 2018.
48. Keith Bradsher, 'Recession elsewhere, but It's booming in China', *The New York Times*, December 9, 2009.
49. Gaia Vince, 'The heat is on over the climate crisis. Only radical measures will work', *Guardian*, May 18, 2019.
50. Amit Katwala, 'The spiralling environmental cost of our lithium battery addiction', *Wired*, August 5, 2018.

Curing Cancer

1. Maayan Jaffe-Hoffman, 'A cure for cancer? Israeli scientists may have found one', *The Jerusalem Post*, January 28, 2019; Ben Feuerherd, 'We'll have a cure for cancer within a year, scientists claim', January 28, 2019.
2. Victoria Forster, 'An Israeli company claims that they will have a cure for cancer in a year. Don't believe them', *Forbes*, January 31, 2019.
3. Cancer Research UK guidance on how long it takes to carry a drug through clinical trials, accessed December 2019, can be found at cancer-researchuk.org/find-a-clinical-trial/how-clinical-trials-are-planned-and-organised/how-long-it-takes-for-a-new-drug-to-go-through-clinical-trials.
4. Matthew Abola & Vinay Prasad, 'The use of superlatives in cancer research', *JAMA Oncol.*, 2016, 2(1), 139-141. DOI: 10.1001/jamaoncol.2015.3931.
5. A. S. Ahmad *et al.*, 'Trends in the lifetime risk of developing cancer in Great Britain: comparison of risk for those born from 1930 to 1960', *Br J Cancer*, 2015, 112(5), 943–947. DOI: 10.1038/bjc.2014.606.
6. Taken from the World Health Organization's fact sheet on cancer globally, August 12, 2018, accessed December 2019, can be found at who.int/news-room/fact-sheets/detail/cancer.
7. Taken from the National Institute of Health's National Cancer Institiute 'About cancer' resource, accessed December 2019, can be found at cancer.gov/about-cancer/understanding/what-is-cancer.

8. Oscar M. Rueda *et al.*, 'Dynamics of breast-cancer relapse reveal late-recurring ER-positive genomic subgroups', *Nature*, 2019, 567(7748), 399-404. DOI: 10.1038/s41586-019-1007-8.

9. Taken from Medscape's guidance on CML, by Emmanuel C Besa *et al.*, updated October 23, 2019, can be accessed at emedicine.medscape.com/article/199425-overview; also from the National Cancer Institute's guidance on Gleevec, accessed December 2019, can be found at cancer.gov/research/progress/discovery/gleevec.

10. *Ibid* (National Cancer Institute).

11. Victoria Forster, 'Surviving cancer: how big data is helping patients live longer, healthier lives', London School of Hygiene & Tropical Medicine, blog post from June 2019.

12. Taken from National Institute of Health's U.S. National Library of Medicine resource tracking clinical trials, accessed July 2019, my search terms and results can be found at clinicaltrials.gov/ct2/results?cond=cancer&term=immunotherapy&cntry=&state=&city=&dist= (there will most likely be far more than 3,000 trials by the time this book is published).

13. Nathan Gay & Vinay Prasad, 'Few people actually benefit from 'breakthrough' cancer immunotherapy', *Stat News*, March 8, 2017.

14. Sharon Begley, 'Beware the hype: Top scientists cautious about fighting cancer with immunotherapy', *Stat News*, September 25, 2016; Sharon Begley, 'Precision medicine, linked to DNA, still too often misses', *Boston Globe*, August 29, 2015.

15. Matthew Michaels, 'The price of a 30-second Super Bowl ad has exploded – but it may be worth it for companies', *Business Insider*, January 25, 2018.

16. The 2015 Opdivo ad, Ad ID: 1260829, accessed December 2019, archived at ispot.tv/ad/AL_Z/opdivo-longer-life.

17. Michael Joyce, 'Consumer drug ads: The harms that come with pitching lifestyle over information', *HealthNewsReview.org*, May 23, 2018.

18. Liz Szabo, 'Cancer treatment hype gives false hope to many patients', *USA Today*, April 27, 2017.

19. Ben Hirschlerl, 'Bristol's drug pricing under fire as UK agency rejects Opdivo', *Reuters*, December 16, 2015.

20. Jon Swallen, 'Drug advertising booms to $6.4 billion', *Kantar US Insights*, May 8, 2017.

21. *See note 17* (Joyce).

22. Michael S. Wilkes *et al.*, 'Direct-to-consumer prescription drug advertising: Trends, impact, and implications', Health Affairs, 2000, 19. DOI: 10.1377/hlthaff.19.2.110; Abby Alpert *et al.*, 'Prescription Drug Advertising and Drug

Utilization: The Role of Medicare Part D', National Bureau of Economic Research, 2015, Working Paper No. 21714. DOI: 10.3386/w21714.

23. 'World first use of gene-edited immune cells to treat "incurable" leukaemia', NHS Foundation Trust, Great Ormond Street Hospital for Children, press release November 5, 2015.

24. John Carroll, 'Months after another lethal setback, Juno finally opts to kill lead CAR-T', *Endpoints News*, March 1, 2017; Clara Rodriguez Fernandez, 'The FDA approves a second CAR-T therapy, cheaper than Novartis', *Labiotech*, October 19, 2017.

25. Michelle Cortez, 'There is a magic bullet for some cancers. What if it misses?', *Bloomberg*, December 3, 2018.

26. Clara Rodriguez Fernandez, 'A cure for cancer? How CAR T-cell therapy is revolutionizing oncology', *Labiotech*, September 10, 2019.

27. Liz Szabo, 'Cascade of costs could push CAR-T therapy to $1.5M per patient', *Endpoints News*, October 17, 2017.

28. Products such as Cellectis' UCART, cellectis.com/en/products/ucarts/; Victoria Forster, 'New experimental therapy CAR NKT-Cells tested in cancer patients or the first time', *Forbes*, May 1, 2019.

29. Brad Plumer *et al.*, 'A simple guide to CRISPR, one of the biggest science stories of the decade', *Vox*, December 27, 2018.

30. Michael Kosicki *et al.*, 'Repair of double-strand breaks induced by CRISPR–Cas9 leads to large deletions and complex rearrangements', *Nat Biotechnology*, 2018, 36, 765–771. DOI: 10.1038/nbt.4192.

31. Sharon Begley, 'A serious new hurdle for CRISPR: Edited cells might cause cancer, two studies find', *Stat News*, June 11, 2018.

32. Victoria Forster, 'Why Do Only Eight Percent Of Cancer Patients In The U.S. Participate In Clinical Trials?' *Forbes*, February 19, 2019.

33. Vinay Prasad, 'FDA's genomic screening approval blurs research and practice', *Medscape*, June 28, 2018.

34. Shraddha Chakradhar, 'It's just in mice! This scientist is calling out hype in science reporting', *Stat*, April 15, 2019.

35. *See note 4* (Abola).

36. *Ibid.*

37. Peter Schmid, 'Atezolizumab and Nab-Paclitaxel in advanced triple-negative breast cancer', *New England Journal of Medicine*, 2018, 379, 2108-2121. DOI: 10.1056/NEJMoa1809615.

38. Taken from the U.S. Department of Health & Human Services Centers for Disease Control and Prevention guidance on TNBC, accessed July 2019, can be found at cdc.gov/cancer/breast/triple-negative.htm.

39. Ned Pagliarulo, 'Benefit of Roche's Tecentriq looks limited in breast cancer', *BioPharma Dive*, October 20, 2018.

40. Kathy D. Miller, 'Kathy Miller on IMpassion130: Immunotherapy a new standard in breast cancer? Not yet', *Medscape*, November 8, 2018.
41. Richard Staines, 'Roche's Tecentriq gets fast review in triple negative breast cancer', *Pharmaphorum*, November 13, 2018; Nick Mulcahy, 'FDA approves first immunotherapy for breast cancer', *Medscape*, March 8, 2019.
42. *See note 39* (Pagliarulo).
43. Andrew McConaghie, 'Keytruda monotherapy fails in triple negative breast cancer', *PM LiVE*, May 21, 2019.
44. Mary Chris Jaklevic, '"Simply cruel": Patient advocates condemn breast cancer immunotherapy hype', *HealthNewsReview.org*, October 24, 2018.
45. Chul Kim & Vinay Prasad, 'Cancer drugs approved on the basis of a surrogate end point and subsequent overall survival: an analysis of 5 years of US Food and Drug Administration approvals', *JAMA Internal Medicine*, 2015, 175(12), 1992-4. DOI: 10.1001/jamainternmed.2015.5868.
46. Chul Kim & Vinay Prasad, 'Strength of validation for surrogate end points used in the US Food and Drug Administration's approval of oncology drugs', *Mayo Clinic Proceedings*, 2016, 91(6), 713–25. DOI: 10.1016/j.mayocp.2016.02.012; Vinay Prasad, 'Calling out the FDA's "We don't regulate medicine" mantra', *Medscape*, May 24, 2018.
47. This is a great piece by the associate editor of the *British Medical Journal*, which I recommend reading in full. Deborah Cohen, 'Cancer drugs: high price, uncertain value', BMJ, 2017, 359. DOI: 10.1136/bmj.j4543.
48. Vinay Prasad *et al.*, 'Low-value approvals and high prices might incentivize ineffective drug development', *Nature Reviews Clinical Oncology*, 2018, 15(7), 399-400. DOI: 10.1038/s41571-018-0030-2.
49. T Fojo *et al.*, 'Unintended consequences of expensive cancer therapeutics – the pursuit of marginal indications and a me-too mentality that stifles innovation and creativity: The John Conley Lecture', *JAMA Otolaryngology Head Neck Surgery*, 2014, 140(12), 1225-36. DOI: 10.1001/jamaoto.2014.1570; Vinay Prasad, 'Why a cancer "moonshot" is unlikely to find us a cure', *The Washington Post*, January 29, 2016; Vinay Prasad, '"Twitter Cheerleaders": A Megaphone for Bad Science', *Medscape*, September 19, 2018.
50. *See note 46* (Kim & Prasad).
51. *See note 48* (Prasad *et al.*).
52. Ajay Aggarwal, 'Demand cancer drugs that truly help patients', *Nature*, 2018, 556(151). DOI: 10.1038/d41586-018-04154-9; Courtney Davis & John Abraham, 'Desperately seeking cancer drugs: explaining the emergence and outcomes of accelerated pharmaceutical regulation', *Sociology*

of Health & Illness, 2011, 33(5), 731-747. DOI: 10.1111/j.1467-9566.2010.01310.x.

53. Nick Triggle, 'Cancer Drugs Fund "huge waste of money"', BBC, April 28, 2017.
54. 'England's Cancer Drugs Fund "Failed to Deliver Meaningful Value to Patients and Society"', *European Society for Medical Oncology*, Annals of Oncology Press Release, April 28, 2017.
55. *See note 47* (Cohen).
56. Ravinder Gabble & Jillian Clare Kohler, 'To patent or not to patent? The case of Novartis' cancer drug Glivec in India', *Global Health*, 2014, 10(3). DOI: 10.1186/1744-8603-10-3.
57. Kaustubh Kulkarni & Suchitra Mohanty, 'Novartis loses landmark India cancer drug patent case', *Reuters*, April 1, 2013; Roger Collier, 'Drug patents: the evergreening problem', *CMAJ*, 2013, 185(9), E385-E386. DOI: 10.1503/cmaj.109-4466.
58. *See note 56* (Gabble & Kohler).
59. *Ibid.*
60. Erik Sherman, 'Drugmakers blame middlemen for high prices but still make huge profits', *Fortune*, February 5, 2019.
61. *See note 47* (Cohen).
62. Brian Resnick, 'Hyped-up science erodes trust. Here's how researchers can fight back.', *Vox*, June 11, 2019.
63. Elena Semino *et al.*, 'The online use of Violence and Journey metaphors by patients with cancer, as compared with health professionals: a mixed methods study', *BMJ Supportive & Palliative Care*, 2017, 7(1), 60-66. DOI: 10.1136/bmjspcare-2014-000785.
64. Kate Granger, 'Having cancer is not a fight or a battle', *Guardian*, April 25, 2014.

LIMITLESS ENERGY

1. 'How the "Zeta fiasco" pulled fusion out of secrecy', ITER, Jan 29, 2018.
2. Roland Pease, *'The story of "Britain's Sputnik"'*, BBC, January 15, 2008.
3. Taken from the United Kingdom Atomic Energy Authority Culham Centre for Fusion Energy's introduction to fusion, accessed December 2019, which can be found at ccfe.ac.uk/introduction.aspx.
4. Joaquin Sánchez, 'Nuclear fusion as a massive, clean, and inexhaustible energy source for the second half of the century: brief history, status, and

perspective', *Energy Science and Engineering*, 2014, 2(4), 165-176. DOI: 10.1002/ese3.43.

5. Ron Dagani, 'Nuclear fusion: Utah findings raise hopes, doubts', *Chemical & Engineering News*, 1989, 67, 4-6; David Voss, 'Whatever happened to cold fusion?', *Physics World*, March 1, 1999.

6. Stephen K Ritter, 'Cold fusion died 25 years ago, but the research lives on', *Chemical Engineering News*, November 7, 2016, 94, 34-39.

7. Robert Arnoux, 'Proyecto Huemul: the prank that started it all', ITER, October 26, 2011.

8. *See note 7* (Arnoux).

9. ITER Communication, 'It's now official: First Plasma in December 2025', ITER, June 20, 2016.

10. John Greenwald, 'Celebrating Lyman Spitzer, the father of PPPL and the Hubble Space Telescope', Princeton Plasma Physics Laboratory, October 23, 2014.

11. Robert Arnoux, 'August 1968: A revolution in fusion', ITER, August 4, 2008.

12. Taken from ITER's introduction to tokamaks, accessed December 2019, and can be found at iter.org/mach/tokamak.

13. Paul Rincon, 'UK centre to shoot for nuclear fusion record', BBC, April 24, 2014.

14. *See note 2* (Pease).

15. *See note 4* (Sánchez).

16. 'On the road to ITER: Milestones', ITER, accessed December 2019, can be found at iter.org/proj/itermilestones#24.

17. Tom Murphy, 'Nuclear fusion', *Do The Math UCSD*, January 31, 2012.

18. *See note 4* (Sánchez).

19. *See note 16* (ITER).

20. 'Fusion furore', *Nature*, 2014, 511, 383-384. DOI: 10.1038/511383b.

21. *See note 16* (ITER).

22. Robert L Hirsch, 'Fusion research: Time to set a new path', *Issues in Science and Technology*, Summer 2015.

23. J Kaslow *et al.*, 'Criteria for practical fusion power systems: report from the EPRI Fusion Panel, Journal of Fusion Energy', *Journal of Fusion Energy*, 1994, 13(2-3), 181-183. DOI: 10.1007/BF02213958.

24. *See note 22* (Hirsch).

25. *Ibid.*

26. *See note 17* (Murphy); Brian Bergstein, 'Finally, fusion power is about to become a reality', *OneZero*, January 3, 2019.

27. Daniel Jassby, 'Fusion reactors: Not what they're cracked up to be', *The Bulletin*, April 19, 2017.

28. 'Fuelling the fusion reactor', ITER, accessed December 2019, can be found at iter.org/sci/FusionFuels; Laila A El-Guebaly *et al.*, 'Goals, challenges, and successes of managing fusion activated materials', *Fusion Engineering and Design*, 2008, 83(7-9), 928-935. DOI: 10.1016/j. fusengdes.2008.05.025.

29. *See note 4* (Sánchez).

30. Taken from the documentary film 'Let there be light', directed by Mila Aung-Thwin & Van Royko, 2017. A great exploration of fusion development I recommend watching.

31. *See note 23* (Kaslow).

32. Robert Arnoux, 'Twist and fuse', ITER, November 2, 2015.

33. Isabella Milch, 'Wendelstein 7-X fusion device produces its first hydrogen plasma', Max Planck Institute for Plasma Physics, February 3, 2016.

34. *See note 4* (Sánchez).

35. *Ibid*; Omar Hurricane, 'Fuel gain exceeding unity in an inertially confined fusion implosion', *Nature*, 2014, 506(7488), 343-348. DOI: 10.1038/nature13008.

36. 'About NIF & photon science', National Ignition Facility & Photon Science, lasers.llnl.gov/about; 'Le Laser Mégajoule', CEA Direction des Applications Militaires, www-lmj.cea.fr/en/lmj/index.htm. Accessed both December 2019.

37. Mihály Heder, 'From NASA to EU: the evolution of the TRL scale in Public Sector Innovation', *The Innovation Journal: The Public Sector Innovation Journal*, 2017, 22, 2017.

38. Jonathan Frochtzwajg, 'The secretive, billionaire-backed plans to harness fusion', *BBC Future*, April 28, 2016.

39. Ray Rothrock, 'What's the big idea?', *Issues in Science and Technology, Winter 2016*; M Mitchell Waldrop, 'Plasma physics: The fusion upstarts', *Nature*, July 24, 2014, 511, 398-400. DOI: 10.1038/511398a; Akshat Rathi, 'In search of clean energy, investments in nuclear-fusion startups are heating up', *Quartz*, September 26, 2018.

40. *Ibid* (Waldrop); *See note 38* (Frochtzwajg); *See note 30* (Aung-Thwin & Royko).

41. *See note 26* (Bergstein); *See note 39* (Rathi).

42. *Ibid* (Rathi).

43. Dick Ahlstrom, 'Chernobyl anniversary: The disputed casualty figures', *The Irish Times*, April 2, 2016.

44. *See note 27* (Jassby); Martin B. Kalinowski & Lars C. Colschen 'International control of Tritium to prevent horizontal proliferation and to foster nuclear disarmament', *Science & Global Security*, 1995, 5(2), 131-203. DOI: 10.1080/08929889508426422.

COMMERCIALISING SPACE

1. 'Review of the space program: Hearings before the Committee on Science and Astronautics, U. S. House of Representatives, Eighty-sixth Congress, second session' Washington, U.S. Government Printing Office, 1960, No. 3 Part 1, 39.
2. Margaret Kane, 'eBay picks up PayPal for $1.5 billion', CNET, August 18, 2002.
3. Carl Sagan, *Pale Blue Dot: A Vision for the Human Future in Space*, Random House, New York, 1994.
4. *Commercial Space Launch Act* (1984), Public law 98-575, 98th Congress.
5. Chris Leadbeater, 'Infinity and beyond: Will Virgin Galactic ever make it into space?', *The Telegraph*, January 18, 2018.
6. Jon Ungoed-Thomas, 'The $80m Virginauts stranded on Earth', *The Times*, September 14, 2014.
7. Erin Durkin, 'Virgin Galactic launches SpaceShipTwo to the edge of space', *Guardian*, December 13, 2018; Jonathan O'Callaghan, '2019 is the year that space tourism finally becomes a reality. No, really', *Wired*, January 24, 2019.
8. Samantha Masunaga, 'Lost in space: They paid $100,000 to ride on Xcor's space plane. Now they want their money back', *Los Angeles Times*, December 30, 2018.
9. Jonathan O'Callaghan, 'Goodbye Mars One, the fake mission To Mars that fooled the world', *Forbes*, February 11, 2019.
10. *Communications Satellite Communications Act of 1962* (1962), Public law 87-624, 87th Congress.
11. From the NASA Space Science Data Coordinated Archive information on Telstar 1, nssdc.gsfc.nasa.gov/nmc/spacecraft/display.action?id=1962-029A.
12. Michael Sheetz, 'Amazon wants to launch thousands of satellites so it can offer broadband internet from space', CNBC, April 4, 2019; Ashlee Vance, 'OneWeb raises fresh $1.25 billion for internet system from space', *Bloomberg*, March 18, 2019.
13. The European Space Agency keeps a tally of the number of debris objects in space at esa.int/Safety_Security/Space_Debris/Space_debris_by_the_numbers, accessed December 2019.
14. Kazunori Takahashi et al., 'Demonstrating a new technology for space debris removal using a bi-directional plasma thruster', *Scientific Reports*, 2018, 8, 14417. DOI: 10.1038/s41598-018-32697-4.
15. 'ESA's e.Deorbit debris removal mission reborn as servicing vehicle',

European Space Agency, December 12, 2018; Astroscale, astroscale.com/about, accessed December 2019.

16. Caleb Henry, 'FCC fines Swarm $900,000 for unauthorized smallsat launch', *SpaceNews*, December 20, 2018.

17. Alan Boyle, 'After satellite flap, Swarm Technologies raises $25M for space-based IoT network', *GeekWire*, January 26, 2019.

18. Bhavya Lal *et al.*, 'Global trends in Space Situational Awareness (SSA) and Space Traffic Management (STM)', IDA Science and Technology Policy Institute, April, 2018, ida.org/-/media/feature/publications/g/gl/global-trends-in-space-situational-awareness-ssa-and-space-traffic-management-stm/d-9074.ashx.

19. For a deep dive into liability for space debris collisions, I recommend the following: Scott Kerr, 'Liability for space debris collisions and the Kessler Syndrome', *The Space Review*, SpaceNews, December 11, 2018 (part 1), December 18, 2018 (part 2).

20. 'Text of the convention for the safety of life at sea', International Conference on Safety of Life at Sea, London, January 20, 1914.

21. 'Treaty on Principles Governing the Activities of States in the Exploration and Use of Outer Space, including the Moon and Other Celestial Bodies', United Nations Office for Disarmament Affairs, opened for signature at London, Moscow & Washington, January 27, 1967. For full details on which countries signed when (some only joined in the 2000s), details can be accessed at disarmament.un.org/treaties/t/outer_space.

22. There are various different approaches to assessing the effectiveness of these bodies, explored in the following fascinating pieces: Edwin W. Paxson III, 'Sharing the Benefits of Outer Space Exploration: Space Law and Economic Development', *Michigan Journal of International Law*, 1993, 14, 509-10; Jon Copley, 'Shedding some light on the International Seabed Authority', University of Southampton 'Exploring Our Oceans' MOOC, March 9, 2014; Leslie Hook & Benedict Mander, 'The fight to own Antarctica', *Financial Times*, May 24, 2018; B. E. Heim, 'Exploring the last frontiers for mineral resources: a comparison of international law regarding the deep seabed, outer space, and Antarctica', *Vanderbilt Journal of Transnational Law*, 1990, 23, 819; D. A. Wirth, 'Public participation in international processes: environmental case studies at the national and international levels', *Colorado Journal of International Environmental Law and Policy*, 1996, 7.

23. J. Ezor, 'Costs overhead: Tonga's claiming of sixteen geostationary orbital sites and the implications for US space policy', *Law and Policy International Business*, 1993, 24.

24. Jeff Foust, 'Luxembourg establishes space agency and new fund', *SpaceNews*, September 13, 2018.

25. Andrew Silver, 'Luxembourg passes first EU space mining law. One can possess the Spice', *The Register*, July 14, 2017.

26. Atossa Araxia Abrahamian, 'How a tax haven is leading the race to privatise space', *Guardian*, September 15, 2017.

27. Some of my thinking in this section was informed by this fascinating essay, and the collection in which it was published: Robert L. Pfaltzgraff, Jr., 'International Relations Theory and Spacepower', published in *Toward a theory of spacepower: Selected essays*, Institute for National Strategic Studies, National Defense University, edited by Charles Lutes, Peter Hays, Vincent Manzo, Lisa Yambrick & Elaine Bunn, *Progressive Management*, January, 2013, 29-43.

28. *See note 21* (UN).

29. Max Mutschler, *Arms control in space: Exploring conditions for preventive arms control*, London, Palgrave Macmillan UK, 2013, 142.

30. For more on this topic, I recommend this dissertation extract: 'IR theory and space: Realism, liberalism and constructivism', *CosmoPolicy*, January 6, 2016.

31. Andrew Wong, 'Space mining could become a real thing – and it could be worth trillions', CNBC, May 15, 2018.

32. 'Asterank: Asteroid and Database Mining Rankings' is a great website resource which tracks over 600,000 asteroids in terms of their mass and composition, and from this, estimates their economic value. It has a 3D visualisation which I got lost in for some time. Accessed December 2016, and can be found at asterank.com.

33. Rachel Riederer, 'Silicon Valley says space mining is awesome and will change life on Earth. That's only half right', *New Republic*, May 20, 2014.

34. Susan Caminiti, 'The billionaire's race to harness the moon's resources', CNBC, April 3, 2014.

35. Charles W. Dunnill & Robert Phillips, 'Making space rocket fuel from water could drive a power revolution on Earth', *The Conversation*, September 27, 2016.

36. Moon Village Association, accessed December 2019, moonvillageassociation.org; 'Deep Space Gateway to open opportunities for distant destinations', NASA, March 28, 2017.

37. 'China to build scientific research station on Moon's south pole', *Xinhua Net*, April 24, 2019; 'After Israel's failed Moon mission, ISRO treads cautious path; postpones Chandrayaan launch to July', *The Economic Times*, April 25, 2019; Laura Parker, 'Beyond Prime: Inside

the race to deliver shipments to the moon', *OneZero*, April 17, 2019; accessed December 2019, ispace-inc.com/aboutus/.

38. Josephine McDermott, 'The archaeological legacy of the Crossrail excavations', BBC, February 10, 2017.

39. In case you want to learn more: Nathan J, Robinson, 'A quick reminder of why colonialism was bad', *Current Affairs*, September 14, 2017.

40. An interesting read which added to my thinking in this section: Fraser MacDonald, 'Anti-Astropolitik: Outer space and the orbit of geography', *Prog Hum Geogr*, 2007, 31. DOI: 10.1177/0309132507081492.

41. Tim Flannery, 'The Great Barrier Reef and the coal mine that could kill it', *Guardian*, August 1, 2014.

42. *See note 21* (UN).

43. There's much debate about the relationship between technology and the widening wealth gap. A few perspectives to consider: Jonathan Mijs, 'The gulf between the rich and poor is widening – we just can't see it', *The Independent*, January 28, 2019; Zia Qureshi, 'Globalization, technology, and inequality: It's the policies, stupid', *Brookings*, February 16, 2018; Larry Elliot, 'Rising inequality threatens world economy, says WEF', *Guardian*, January 11, 2017; Christoffer Hernæs, 'Is technology contributing to increased inequality?', *TechCrunch*, March 29, 2017; Katie Allen, 'Big Tech's big problem – its role in rising inequality', *Guardian*, August 2, 2015; Florence Jaumotte *et al.*, 'IMF Survey: Technology Widening Rich-Poor Gap', International Monetary Fund report summary, October 10, 2007.

Understanding the Universe

1. SXSW 2018 event statistics can be accessed here explore.sxsw.com/ hubfs/Hosted%20Files/2018-SXSW-Event-Stats-April-2018.pdf.

2. Whurley's SXSW keynote youtube.com/watch?v=3tNlAtQz17Q, accessed December 2019.

3. T Kleinjung *et al.*, 'Factorization of a 768-Bit RSA Modulus'. Chapter in: T Rabin, 'Advances in Cryptology – CRYPTO 2010', 2010, 6223. DOI: 10.1007/978-3-642-14623-7_18.

4. 'Progress in general purpose factoring', *AI Impacts*. March 16, 2016.

5. Katherine Bourzac, 'Chemistry is quantum computing's killer app', *Chemical Engineering News*, October 30, 2017, 95, 27-31.

6. Markus Reiher *et al.*, 'Elucidating reaction mechanisms on quantum computers', *PNAS*, 2017, 114, 7555-7560. DOI: 10.1073/pnas.1619152114.

7. Thomas Sullivan, 'A tough road: cost to develop one new drug is $2.6

billion; approval rate for drugs entering clinical development is less than 12%', *Policy & Medicine*, March 21, 2019.

8. 'NIST reveals 26 algorithms advancing to the Post-Quantum Crypto "semifinals"', *NIST*, January 30, 2019.

9. Peter Shor, 'Algorithms for quantum computation: discrete logarithms and factoring', IEEE – Proceedings 35th Annual Symposium on Foundations of Computer Science, 1994. DOI: 10.1109/SFCS.1994.365700.

10. Ashley Montanaro, 'Quantum algorithms: an overview', *npj Quantum Information*, 2016, 2, 15023. DOI: 10.1038/npjqi.2015.23.

11. Richard Feynman, 'Simulating physics with computers', *International Journal of Theoretical Physics*, 1982, 21(6-7), 467-488. DOI: 10.1007/BF02650179.

12. Katia Moskvitch, 'The argument against quantum computers', *Quanta*, February 7, 2018.

13. Katia Moskvitch, 'Inside the high-stakes race to make quantum computers work', *Wired*, March 6, 2019.

14. Craig Gidney & Martin Ekerå, 'How to factor 2048 bit RSA integers in 8 hours using 20 million noisy qubits', 2019, arXiv:1905.09749 [quant-ph].

15. Will Knight, 'IBM raises the bar with a 50-qubit quantum computer', *MIT Tech Review*, November 10, 2017; Tristan Greene, 'Google reclaims quantum computer crown with 72 qubit processor', *The Next Web*, March 6, 2018.

16. John Preskill, 'Quantum Computing in the NISQ era and beyond', *Quantum*, 2018, 2, 79.

17. Nicola Jones, 'Computing: The quantum company', *Nature*, 2013, 498, 286-288. DOI: 10.1038/498286a.

18. Erico Guizzo, 'Loser: D-Wave does not quantum compute', *IEEE Spectrum*, December 31, 2009.

19. Sebastian Anthony, 'First ever commercial quantum computer now available for $10 million', *ExtremeTech*, May 20, 2011.

20. Stephanie Mlot, 'Quantum computing firm gets $30M boost from Bezos, CIA', *PCmag*, October 5, 2012.

21. Robert McMillan, 'Google, NASA Open New Lab to Kick Tires on Quantum Computer', *Wired*, May 16, 2013.

22. Quentin Hardy, 'A quantum computer aces its test', *The New York Times*, May 8, 2013; Terrence O'Brien, 'Google and NASA team up for D-Wave-powered Quantum Artificial Intelligence Lab', *Engadget*, May 16, 2013; Jeremy Hsu, 'D-Wave's quantum computing claim gets boost in testing', *IEEE Spectrum*, May 10, 2013.

23. Troels F Rønnow *et al.*, 'Defining and detecting quantum speedup', *Science*, 2014, 345(6195), 420-424. DOI: 10.1126/science.1252319.

24. Scott Aaronson, 'D-Wave: Truth finally starts to emerge', *Shtetl-Optimized*, May 16, 2013; Paul Rincon, 'D-Wave: Is $15m machine a glimpse of future computing?', BBC, May 20, 2014; Russell Brandom, 'Google's quantum computer just flunked its first big test', *The Verge*, June 19, 2014.

25. Emily Reynolds, 'Google's quantum computer is 100 million times faster than your PC', *Wired*, December 9, 2015; Iain Thomson, 'Google says its quantum computer is 100 million times faster than PC', *The Register*, December 9, 2015.

26. Larry Hardesty, '3Q: Scott Aaronson on Google's new quantum-computing paper', *MIT News*, December 11, 2015.

27. James McLeod, 'D-Wave's quantum computers "can save energy, solve climate change"', *Financial Post*, December 20, 2018.

28. *Ibid.*

29. Emil Protalinski, 'D-Wave previews quantum computing platform with over 5,000 qubits', *Venture Beat*, February 27, 2019.

30. Ryan F Mandelbaum, 'Why Did NASA, Lockheed Martin, and Others Spend Millions on This Quantum Computer?', *Gizmodo*, January 1, 2019; Mark Harris, 'Ford Signs Up to Use NASA's Quantum Computers', *IEEE Spectrum*, January 17, 2019.

31. Vivek Wadhwa, 'Quantum computers may be more of an imminent threat than AI', *The Washington Post*, February 5, 2018.

32. Daniel Crevier, *AI: the tumultuous history of the search for artificial intelligence*, New York, Basic Books, 1993, 203.

33. James Vincent, 'Forty percent of "AI startups" in Europe don't actually use AI, claims report', *The Verge*, March 5, 2019.

34. Philipp Gerbert & Frank Rueß, 'The next decade in quantum computing – and how to play', BCG, November 15, 2018; for a comprehensive list of venture capital players in quantum computing see the Quantum Computing Report site quantumcomputingreport.com/players/venture-capital/; Jeff Vance, '10 hot quantum-computing startups to watch', *Insider Pro*, February 15, 2019; 'The race is on to dominate quantum computing', *The Economist*, August 18, 2018.

35. Angus Loten, 'U.S. government urged to boost technology R&D', *The Wall Street Journal*, May 22, 2019; Jonathan Gruber & Simon Johnson, *Jump-starting America: How breakthrough science can revive economic growth and the American dream*, 1st edition, 2019, PublicAffairs.

36. *See note 11* (Feynman).

JUST THINK IT

1. Tim Urban, 'Neuralink and the brain's magical future', *Wait But Why*, November 20, 2017.

2. 'Elon Musk wants to connect brains to computers with new company', *Guardian*, March 28, 2017; Darrell Etherington, 'Elon Musk's Neuralink wants to boost the brain to keep up with AI', *TechCrunch*, March 27, 2017; Aatif Sulleyman, 'Elon Musk to plant computers in human brains to prevent AI robot uprising', *The Independent*, March 28, 2017.

3. Jacques Vidal, 'Toward direct brain-computer communication', *Annual Review of Biophysics and Bioengineering*, 1973, 2, 157-180. DOI: 10.1146/annurev.bb.02.060173.001105.

4. P. R. Kennedy & R. A. E. Bakay, 'Restoration of neural output from a paralyzed patient by a direct brain connection', *NeuroReport*, 1998, 9(8), 1707–11. DOI: 10.1097/00001756-199806010-00007; Adam Piore, 'To study the brain, a doctor puts himself under the knife', *MIT Technology Review*, November 9, 2015.

5. Taken from BrainGate about page, accessed July 2019, can be found at braingate.org/about-braingate/.

6. Taken from National Institute of Health's Brain Initiative funding information for cleared initiatives, accessed July 2019, can be found at brain-initiative.nih.gov/funding/cleared-initiatives.

7. Human Brain Project, accessed July 2019, can be found at humanbrain-project.eu/en/.

8. 'Paralyzed man feels touch through mind-controlled robot hand', *Stat*, October 13, 2016.

9. Dongjin Seo *et al.*, 'Neural dust: An ultrasonic, low power solution for chronic brain-machine interfaces', 2013, arXiv:1307.2196 [q-bio.NC]; Katherine Bourzac, 'Brain interfaces made of silk', *MIT Technology Review*, April 19, 2010.

10. Sarah Jarvis & Simon R. Schultz, 'Prospects for Optogenetic Augmentation of Brain Function', *Front Syst Neurosci*, 2015, 9. DOI: 10.3389/fnsys.2015.00157.

11. Tian-Ming Fu et al., 'Stable long-term chronic brain mapping at the single-neuron level', *Nature Methods*, 2016, 13(10), 875-882. DOI: 10.1038/nmeth.3969; Jia Liu et al., 'Syringe-injectable electronics', *Nature Nanotechnology*, 2015, 10, 629–636. DOI: 10.1038/nnano.2015.115.

12. Antonio Regalado, 'The thought experiment', *MIT Technology Review*, June 17, 2014.

13. Antonio Regalado, 'Scientists have found a way to decode brain signals into speech', *MIT Technology Review*, April 24, 2019.

14. Margi Murphy, 'Brain-stimulating headset aims to replace sleeping pills', *The Telegraph*, June 26, 2018; dreem.com, accessed July 2019.

15. choosemuse.com/muse-2, accessed July 2019.

16. Chris Taylor, 'Muse 2 review: The world's best meditation tech just got even better', *Mashable*, October 30, 2018.

17. Brian Bergstein, 'Here's what brain-stimulating memory enhancement feels like', *OneZero*, April 17, 2019; Megan Theilking, '"A cavalier approach": Experts urge the companies behind brain wearables to rein in their claims', *Stat*, May 28, 2019; Alice Klein, 'Brain-zap therapy may throw people with depression into a rage', *New Scientist*, October 26, 2017.

18. Laura Steenbergen *et al.*, '"Unfocus" on foc.us: commercial tDCS headset impairs working memory', *Experimental Brain Research*, 2016, 234(3), 637-643. DOI: 10.1007/s00221-015-4391-9; Ephrat Livni, 'The best brain hack for learning faster is one you already know', *Quartz*, February 21, 2019.

19. haloneuro.com, accessed July 2019.

20. Jeff Bercovici, 'I Zapped My Brain With Halo Sport to See If It Would Boost My Athletic Performance', *Men's Health*, December 10, 2018; Alex Hutchinson, 'For the Golden State Warriors, brain zapping could provide an edge', *The New Yorker*, June 15, 2016.

21. Lingyan Huang *et al.*, 'Transcranial Direct Current Stimulation with Halo Sport enhances repeated sprint cycling and cognitive performance', *Frontiers in Physiology*, 2019, 10, 118. DOI: 10.3389/fphys.2019.00118; Seung-Bo Park et al., 'Transcranial Direct Current Stimulation of motor cortex enhances running performance', *PLoS ONE*, 2019, 14(2), e0211902. DOI: 10.1371/journal.pone.0211902.

22. modiushealth.com/pages/science, accessed July 2019.

23. Claudia Tanner, 'Could this headset help you lose weight? Cutting-edge device sends signals to the brain to shed body fat without changes in diet or exercise', *Mail Online*, August 7, 2017.

24. modiushealth.com/pages/research#null, accessed July 2019.

25. I. Coates McCall et al., 'Owning ethical innovation: claims about commercial wearable brain technologies', *Neuron*, 2019, 102(4), 728–31. DOI:10.1016/j.neuron.2019.03.026.

26. Xavier Symons, 'Direct to consumer brain stimulation is booming, but ethical concerns remain', *BioEdge*, March 10, 2019.

27. *See note 1* (Urban).

28. Elon Musk at Recode's Code Conference in 2016, video accessed July 2019 at youtube.com/watch?v=ZrGPuUQsDjo.

29. Tian-Ming Fu *et al.*, 'Stable long-term chronic brain mapping at the single-neuron level', *Nat Methods*, 2016, 13(10), 875–882. DOI:10.1038/nmeth.3969.

30. Angela Chen, 'Elon Musk's dreams of merging AI and brains are likely to remain just that – for at least a decade', *The Verge*, April 21, 2017.

31. Antonio Regalado, 'The entrepreneur with the $100 million plan to link brains to computers', *MIT Technology Review*, March 16, 2017.

32. Antonio Regalado, 'Meet the Guys Who Sold "Neuralink" to Elon Musk without Even Realizing It', *MIT Technology Review*, April 4, 2017.

33. *Ibid.*

34. Anthony Cuthbertson, 'Facebook's mind-reading chief quits secretive building 8 project', *Newsweek*, October 20, 2017.

35. Brian Resnick, 'Facebook's plan to tap into our brains, fact-checked by a neuroscientist', *Vox*, April 20, 2017.

36. Cade Metz, 'Facebook's Race to Link Your Brain to a Computer Might Be Unwinnable', *Wired*, April 27, 2017.

37. From a video Zuckerberg shared on his public Facebook profile, accessed July 2019 at facebook.com/zuck/videos/vb.4/10103661167577621.

38. *See note 35* (Resnick).

39. Noam Cohen, 'Zuckerberg wants Facebook to build a mind-reading machine', *Wired*, March 7, 2019.

40. April Glaser & Kurt Wagner, 'Facebook is developing a way to read your mind', *Vox*, April 19, 2017.

41. Axios interview aired on 'Axios on HBO', excerpts are shared in the 'Axios AM' newsletter sent on November 26, 2018, accessed July 2016 at axios.com/newsletters/axios-am-6a50596b-1f0e-4913-a0b3-76abe8986763.html.

42. Erin Biba, 'What it's like when Elon Musk's Twitter mob comes after you', *Daily Beast*, May 28, 2018.

43. Defense Advanced Research Projects Agency, 'Six Paths to the Nonsurgical Future of Brain-Machine Interfaces', DARPA press release, May 20, 2019.

44. *Ibid.*

45. Stephen Chen, '"Forget the Facebook leak": China is mining data directly from workers' brains on an industrial scale', *South China Morning Post*, April 29, 2018.

46. Ravie Lakshmanan, 'Amazon confirms it retains your Alexa voice recordings indefinitely', *The Next Web*, July 3, 2019.

47. Elon Musk on Axios on HBO, quote taken from video excerpt on YouTube, 'Axios: Merging AI & Man – Elon Musk's Vision for The Future | HBO', November 25, 2018, can be found at youtube.com/watch?v=y6GEugjulPw.

48. *See note 41* (Axios AM).

DECISION-MAKING ON TAP

1. Caroline Criado-Perez, 'The deadly truth about a world built for men – from stab vests to car crashes', *Guardian*, February 23, 2019; C J Kahane, 'Injury vulnerability and effectiveness of occupant protection technologies for older occupants and women', National Highway Traffic Safety Administration, 2013.

2. 'Clinical Trials Have Far Too Little Racial and Ethnic Diversity', *Scientific American*, September 1, 2018.

3. Aylin Caliskan, Joanna Bryson, & Arvind Narayanan, 'Semantics derived automatically from language corpora contain human-like biases', *Science*, 2017, 356 (6334), 183–186. DOI: 10.1126/science.aal4230; Joanna Bryson, 'Three very different sources of bias in AI, and how to fix them', *Adventures in NI*, July 7, 2017.

4. Lauren Kirchner, 'Where traditional DNA testing fails, algorithms take over', *ProPublica*, November 4, 2016.

5. Katherine Kwong, 'The algorithm says you did it: the use of black box algorithms to analyze complex DNA evidence', *Harvard Journal of Law & Technology*, Fall 2017, 31(1), 275-301.

6. Matthew Shaer, 'The false promise of DNA testing', *The Atlantic*, June 2016.

7. Lauren Kirchner, 'Traces of crime: How New York's DNA techniques became tainted', *The New York Times*, September 4, 2017.

8. Christoph Molnar created a great open access self-published guide which I found very useful in this section: Christoph Molnar, 'Interpretable machine learning: a guide for making blackbox models explainable', November 15, 2019, can be found at christophm.github.io/interpretable-ml-book/.

9. Pang Wei Koh & Percy Liang, 'Understanding black-box predictions via influence functions', 2017, arXiv:1703.04730 [stat.ML].

10. Sandra Wachter & Brent Mittelstadt, 'A Right to Reasonable Inferences: Re-Thinking Data Protection Law in the Age of Big Data and AI', *Columbia Business Law Review*, 2019, 2.

11. Sameera Karnik & Amar Kanekar, 'Ethical issues surrounding end-of-life care: A narrative review', *Healthcare*, 2016, 4(2), 24. DOI: 10.3390/healthcare4020024

12. Mary Ann Baily, 'Futility, autonomy, and cost in end-of-life care', *The Journal of Law, Medicine & Ethics*, 2011, 39(2), 172-82. DOI: 10.1111/j.1748-720X.2011.00586.x.

13. Aviva Rutkin, 'People will follow a robot in an emergency – even if it's wrong', *New Scientist*, February 29, 2016.

14. Guy Shefer *et al.*, 'Only making things worse: A qualitative study of the

impact of wrongly removing disability benefits from people with mental illness', *Community Mental Health Journal*, 2016, 52(7), 834–841. DOI: 10.1007/s10597-016-0012-8.

15. Rebecca Greenfield, 'Machines are better than humans at hiring the best employees', *Bloomberg*, November 17, 2015.

16. 'The Future of Jobs Report 2018', *World Economic Forum Centre for the New Economy and Society*, accessed December 2019, can be found at weforum.org/docs/WEF_Future_of_Jobs_2018.pdf.

17. Keiichi Murayama, '"Robot taxes" will help keep humans employed, Bill Gates predicts', *Nikkei Asian Review*, November 3, 2018.

18. More information on Project Debater on the IBM website, accessed July 2019, can be found at research.ibm.com/artificial-intelligence/project-debater.

19. Jonathan Haidt & Craig Joseph, 'Intuitive ethics: how innately prepared intuitions generate culturally variable virtues', *Daedalus*, 2004, 133(4), 55-66. DOI: 10.1162/0011526042365555.

20. The research is explored further in this book: Robert Sapolsky, *Behave: The Biology of Humans at Our Best and Worst*, Penguin Random House, 2017, 450.

21. Vincent Conitzer *et al.*, 'Moral decision-making frameworks for artificial intelligence. In Proceedings of the Thirty-First AAAI Conference on Artificial Intelligence (AAAI'17)', *AAAI Press*, 2017, 4831-4835.

22. Vincent Conitzer et al., 'Moral decisio- making frameworks for artificial intelligence', Proceedings of the Thirty-First AAAI Conference on Artificial Intelligence, 2017, 4831-4835.

FINDING LIFE ELSEWHERE

1. David S. McKay *et al.*, 'Search for past life on Mars: Possible relic biogenic activity in Martian meteorite ALH84001', *Science*, 1996, 273(5277), 924-930. DOI: 10.1126/science.273.5277.924.

2. Matt Crenson, 'Still searching for life on Mars', NBC News, August 6, 2006.

3. 'Pres. Clinton's Remarks on the Possible Discovery of Life on Mars (1996)', Footage from the William J. Clinton Presidential Library, August 7, 1996, can be found on YouTube at youtube.com/watch?v=pHhZQWAtWyQ (accessed July 2019).

4. 'Life on Mars?' *The New York Times*, August 8, 1996.

5. D. C. Golden et al., 'Evidence for exclusively inorganic formation of magnetite in Martian meteorite ALH84001', *American Mineralogist*, 2004, 89, 681–695. DOI: 10.2138/am-2004-5-602.

6. Christopher Hitchens, *God Is Not Great: How Religion Poisons Everything*, Atlantic Books, London, 2007,150.

7. Francis Crick & Leslie Orgel, 'Directed panspermia', *Icarus*, 1973, 19(3), 341–8. DOI: 10.1016/0019-1035(73)90110-3.

8. Francis Crick, *Life Itself: Its Origin and Nature*, Simon & Schuster, New York, 1981.

9. Fred Hoyle & Chandra Wickramasinghe, *Diseases from Space*, J.M. Dent & Sons Ltd., London, 1979.

10. A. J. Charig *et al.*, 'Archaeopteryx is not a forgery', *Science*, 1986, 232(4750), 622–6. DOI: 10.1126/science.232.4750.622.

11. J. Harris *et al.*, 'Detection of living cells in stratospheric samples', Proceedings of SPIE Conference 4495, 2002, 192.

12. Phil Plait, 'Has Alien Life Been Found in Earth's Atmosphere? I'm Gonna Go With "No"', *Slate*, September 20, 2013; Ellie zolfagharifard, 'Scientists claim to have found evidence of alien life: Balloon sent to edge of atmosphere picks up organisms "that can only have come from space", *Mail Online*, September 10, 2013; Richard Gray, 'Alien life found living in Earth's atmosphere, claims scientist', *The Telegraph*, September 19, 2013; Tom Mendelsohn, 'The truth is out there (above Cheshire, that is): British scientists claim to have found proof of alien life', September 19, 2013.

13. Karl Popper, *The Logic of Scientific Discovery*, Routledge, Abingdon-on-Thames, 1959. Popper's original work in German was published in 1934, *Logik der Forschung. Zur Erkenntnistheorie der modernen Naturwissenschaft.*

14. Massimo Pigliucci, *Nonsense on Stilts: How to Tell Science From Bunk*, University of Chicago Press, Chicago, 2010, 1-5.

15. A. I. Goldman, 'Experts: Which Ones Should You Trust?', in E. Selinger & R. P. Crease (eds), *The Philosophy of Expertise*, Columbia University Press, New York, 2006.

16. Sean Carroll, *The Big Picture: On the Origins of Life, Meaning and the Universe Itself*, Oneworld, London, 2016, 80.

17. According to NASA's most recent astrobiology strategy, penned in 2015, accessed July 2019, can be found at nai.nasa.gov/media/medialibrary/2016/04/NASA_Astrobiology_Strategy_2015_FINAL_041216.pdf.

18. Gerda Horneck et al., 'AstRoMap European Astrobiology Roadmap', *Astrobiology*, 2016, 16(3), DOI: 10.1089/ast.2015.1441.

19. Tom Fish, 'Water found on Mars: NASA discovers ancient polar ice cap remnants on Red Planet', *Express*, May 29, 2019; Meghan Bartels, 'Huge amount of water ice is spotted on Mars (it could be long-lost polar ice caps)', *Space.com*, May 23, 2019; Mary Halton, 'Liquid water "lake" revealed on Mars', BBC, July 25, 2018; Josh Davis, 'Liquid water found

beneath the surface of Mars', Natural History Museum, July 26, 2018; NASA, 'NASA confirms evidence that liquid water flows on today's Mars', Press Release 15–195, September 28, 2015; Mike Wall, 'Curiosity rover makes big water discovery in Mars dirt, a "Wow Moment"', *Space.com*, September 26, 2013; NASA, 'NASA rover finds old streambed on Martian surface', Press release 2012-305, September 27, 2012.

20. Original story of the Drake equation is told in this book: Frank Drake & Dava Sobel, *Is anyone out there? The scientific search for extraterrestrial intelligence*, Delacorte Press, New York, 1992; for a more modern take on the Drake equation, reflecting more current views from the field, I'd recommend this paper: Mark J Burchill, 'W(h)ither the Drake equation', *International Journal of Astrobiology*, 2006, 5(3), 243-250. DOI: 10.1017/S1473550406003107.

21. *Ibid* (Drake & Sobel), 55-62; Nicolas Glade, Pascal Ballet, & Olivier Bastien 'A stochastic process approach of the drake equation parameters', *International Journal of Astrobiology*, 2012, 11(2), 103-108. DOI: 10.1017/S1473550411000413.

22. 'SETI at 50'. *Nature*, 2009, 461, 316. DOI: 10.1038/461316a.

23. Jason T Wright & Michael P Oman-Reagan, 'Visions of human futures in space and SETI', *International Journal of Astrobiology*, 2018, 17(2), 177-188. DOI: 10.1017/S1473550417000222.

24. John Noble Wilford, 'Ear to the universe is plugged by budget cutters', *The New York Times*, October 7, 1993; Marina Koren, 'Congress is quietly nudging NASA to look for aliens', May 9, 2018.

25. *See note 23* (Wright & Oman-Reagan).

26. Robert Sanders, 'UC Berkeley passes management of Allen Telescope Array to SRI', *Berkeley News*, April 13, 2012; Breakthrough Initiatives, 'Yuri Milner and Stephen Hawking announce $100 million breakthrough initiative to dramatically accelerate search for intelligent life in the universe', press release, July 20, 2015, accessed July 2019, can be found at breakthroughinitiatives.org/news/1.

27. For a fuller exploration of so-called 'astrobioethics', I recommend checking out this 2018 paper: Octavio A Chon-Torres, 'Astrobioethics', *International Journal of Astrobiology*, 2018, 17(1), 51-56. DOI: 10.1017/S1473550417000064.

28. Steven Johnson, 'Greetings, E.T. (Please Don't Murder Us.)', *The New York Times*, June 28, 2017.

29. The full mission of METI, accessed July 2019, can be found at meti.org/en/mission.

30. The full statement from the group who penned the petition, 'Regarding messaging to extraterrestrial intelligence (METI) / active searches for

extraterrestrial intelligence (active SETI)', accessed July 2019, can be found at setiathome.berkeley.edu/meti_statement_0.html.

31. Rowan Hooper, 'Contacting aliens: War of the worlds or war over cash?', *New Scientist*, February 15, 2015.

32. Seth Shostak, 'Should we keep a low profile in space?', *The New York Times*, March 27, 2015.

33. Bruce Dorminey, 'Hiding from hostile space aliens is no longer an option', *Forbes*, May 19, 2016.

34. Pallab Ghosh, 'Scientists in US are urged to seek contact with aliens', BBC, February 12, 2015.

35. Ian Crawford, 'Widening perspectives: the intellectual and social benefits of astrobiology (regardless of whether extraterrestrial life is discovered or not)', *International Journal of Astrobiology*, 2018, 17(1), 57-60. DOI: 10.1017/S1473550417000088.

36. 'American Foreign Policy: Current Documents', Department of State Publication 8372, Washington DC, 1968, Historical Office, Bureau of Public Affairs, 147.

CONCLUSION

1. H. G. Wells, *The Outline of History*, vol. 2 (London, 1921), p. 594.

Index

Want more from this author?

Gemma Milne shares exclusive writing, curated recommendations, and updates on current and future work in her free weekly newsletter *Brain Reel*.

Get it delivered straight to your inbox by signing up at:
gemmamilne.co.uk/readbrainreel